CALCULUS FUNDAMENTALS EXPLAINED

Samuel Horelick

Copyright © 2021 by Samuel Horelick. All rights reserved.

No part of this book may be reproduced, stored in a retrieval system, or transmitted in any form or by any means, electronic, mechanical, photocopying, recording, scanning, or otherwise without prior written permission of the author, except as permitted under section 107 or 108 of the 1976 United States Copyright Act.

About the author: Dr. Samuel Horelick is mathematics professor and higher education consultant. He has graduated from three Universities with four degrees in Mathematics, Philosophy, Mathematical Education, and Theology.

CONTENTS

Introduction

CHAPTER 1 – FUNCTIONS

What is Function?
Functions and Graphs
Vertical Line Test: is it a Function?
Horizontal Line Test: is it One-to-One Function?
Arithmetic Combinations and Composition of Functions
Inverse Functions
Families of Functions
Symmetry, Even and Odd Functions
Zeroes of a Function

CHAPTER 2 – LIMITS

The Concept and Properties of Limits
Continuity
Infinite Limits and Vertical Asymptotes
Limits at Infinity and Horizontal Asymptotes
Tangent Lines, Areas, and Limits

CHAPTER 3 – DERIVATIVES

Derivative as a Limit
Derivatives and Rules of Differentiation
Writing Equations of Lines Tangent to Function $f(x)$
Average and Instantaneous Velocity
Chain Rule
Implicit Differentiation
Differentials
Increasing and Decreasing Functions, Mean-Value Theorem
First Derivative Test
Second Derivative Test
Concavity and Inflection Points
Absolute Maximum and Minimum, Extreme-Value Theorem
Business Applications

CHAPTER 4 – INTEGRALS

Indefinite Integral: Integration as Anti-differentiation
Integration by Substitution
Initial Conditions and Particular Solutions to Differential Equations
Definite Integral: Area Under the Graph
Area Between Two Graphs
Volume of a Solid of Revolution

INTRODUCTION

This textbook is a product of two decades of teaching Calculus. Over this time I have taught students of every imaginable cultural, social, and educational background, from the affluent communities in South Florida to the inner-city New York, not to mention various colleges and innumerable hours of private tutoring. And regardless of student's background and the level of preparation, most students seem to have one deeply ingrained misconception about Calculus: "Calculus is something so difficult and 'far out' that most people cannot really do it at all." This misconception is so blatantly false, but so widely held, that it is probably the school system itself that propagates this ridiculous, silly notion.

One serious difficulty in learning Calculus today is a lack of simple and clear textbooks. There are dozens of huge, expensive textbooks that are so needlessly complicated as to be practically incomprehensible. It seems that every author of Calculus textbook believes that every student is a born mathematician that wants to become a math professor. They forget that a mark of a good teacher is the ability to make complicated things seem easy and simple. A good textbook must *explain and teach* its subject matter, *helping* the student to learn and say: "That's all? That's easy!"

The effectiveness of this textbook was tested in one of the most challenging (and rewarding) teaching environments – the inner-city high schools of New York. And then it was improved, updated, and implemented to teach Science, Math, and Engineering students in college.

In this textbook you will find all mathematics needed for the first Calculus course; clear explanations of all the principal concepts of Calculus; coverage of all course fundamentals; effective problem-solving skills and strategies; and fully worked out problems with complete, step-by-step explanations and answers.

The purpose of this book is to enable any student to master the fundamentals of Calculus in one semester – four months. The only requirement is an honest desire to learn. The material in this textbook is prepared and sequenced in a particular way to allow for easy, logical progression from the simple to the more elaborate concepts and exercises.

By carefully reading all material, working through all examples and solving all exercises, the student will easily master the fundamentals of Calculus in one semester.

This textbook is intended as the main text for the first Calculus course in College. It may also be used as an explanatory or supplemental text with a different textbook.

Calculus is the mathematics of change, and change is represented by functions. The basic operations in Calculus are differentiating and integrating functions. Whatever we do in Calculus, we deal with functions – we find limits of functions, we differentiate functions, we integrate functions. To understand Calculus students must understand functions first: what sort of relation is called "function", how to graph functions, how to find slope of a line, how to write equations of lines. These are very basic topics, but the understanding of these basic concepts provides a necessary foundation and a starting point in the study of Calculus. This book explains and *teaches* Calculus in a logical, handy and succinct format without overwhelming you with unnecessary details.

The ideas and principles of what eventually became known as Calculus can be traced back to the Ancient Greek mathematicians. These ideas were developed by mathematicians and philosophers through the Middle Ages, but the actual rigorous clarification of the fundamental principles of Calculus was made independently by Isaac Newton in England and Gottfried Leibniz in Germany in seventeenth century. We can simplify the fundamentals into three classes of problems:

1. How to find an equation of the line tangent to the graph of a given function at a given point.

2. How to find the area of a given region.

3. How to find velocity and acceleration at an instant, given a formula for the distance traveled by an object in a specified period of time. Conversely, given a formula for acceleration at a given instant, how to find the distance traveled by this object in a specified period of time. Each of these problems involves *infinite process:*

For the Tangent Line problem, the infinite process is the process of drawing secant lines through two points P and Q on the graph of function f(x) as point Q "moves" infinitely closer and closer to point P so that the distance between P and Q approaches zero.

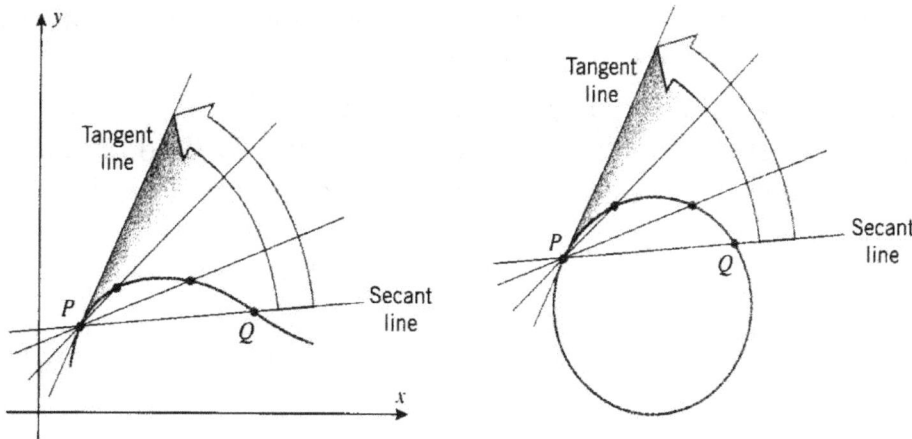

For the Area problem, the infinite process is the measurement of the area under the graph of function using more and more rectangles with progressively smaller widths.

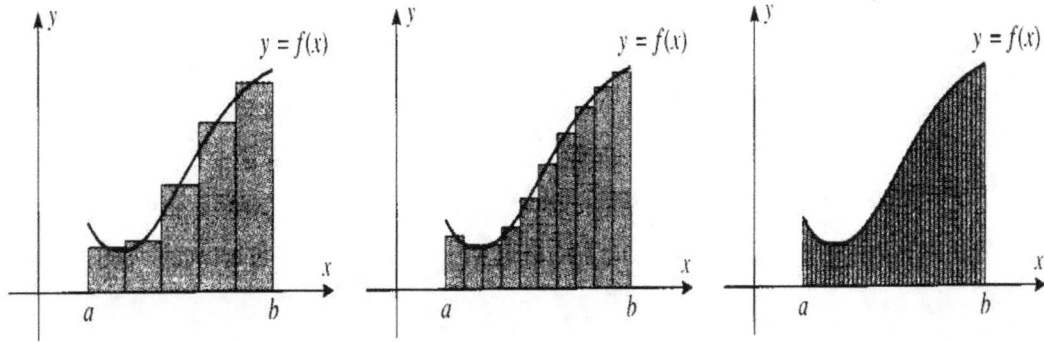

For the Instantaneous Velocity and Acceleration problem, the infinite process is the limiting value of the average speed computed over smaller and smaller time intervals, progressively closer to the specified instant of time t.

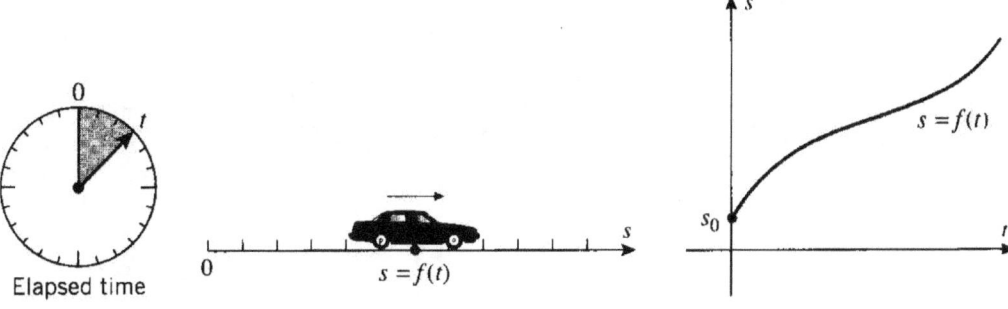

Calculus is a set of ideas that provides a way of analyzing the world around us. As with all mathematics courses, Calculus involves equations and formulas. But even if you learn all the formulas and solve all the equations but do not master the underlying ideas, you would have learned really nothing. But if you carefully study and understand the interrelated ideas of Calculus, not only will you have the tools to go beyond what other people have done, you will also be amazed that Calculus is not really difficult at all.

CHAPTER 1 – FUNCTIONS

WHAT IS FUNCTION?

Function is simply a relation between two sets of objects (or numbers) where each member of the set called "Domain" is assigned to exactly one member of another set called "Range".

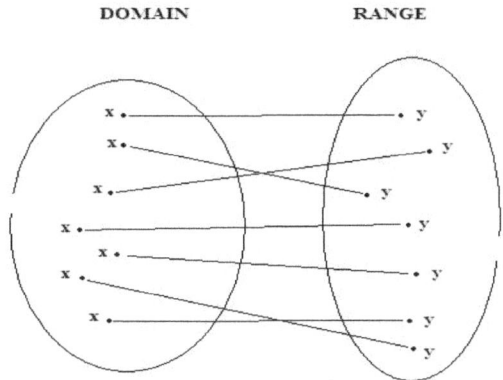

Some functions may assign several members from the Domain to exactly one member of the Range:

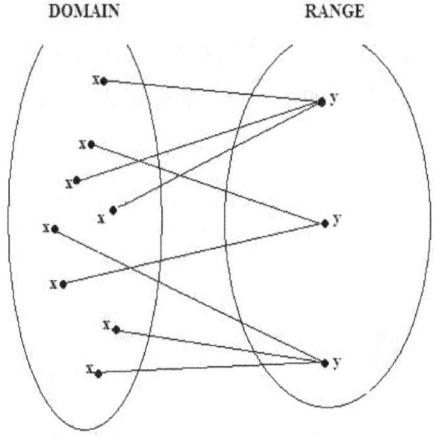

Relation That Is Not function:

However, one object in the Domain may not be assigned to more than one object in the Range. Think of a relation between the set of policemen in a precinct and the set of police cars. If there are enough cars, each policeman can be assigned to one car.

If not, two policemen may be assigned to the same car, or three, etc. What is not possible is to assign one policeman to more than one car at the same time.

A relation that assigns one element of the Domain to more than one element of the Range is not function.

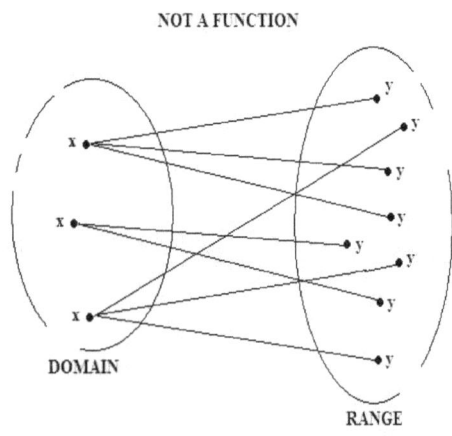

NOT A FUNCTION

DOMAIN RANGE

FUNCTIONS AND GRAPHS:

Example:

Let set D = {2, 4, 6, 8}, let function F assign each element of set D (domain) to an element of set R (range) using the rule: "an element of set D plus 10 is equal to an element of set R". Then,

F(2) = 2 + 10 = 12
F(4) = 4 + 10 = 14
F(6) = 6 + 10 = 16
F(8) = 8 + 10 = 18 and therefore, set R = {12, 14, 16, 18}.

If we denote the numbers in the set D by "x" and denote the numbers in the set R by "y", then this function F is written F(x) = x + 10 or y = x + 10. The meaning of "y" is the same as the meaning of "F(x)" – it is the Range of the function.

Function may be represented by ordered pairs (x, y) or (x, f(x)), where x is an element of the Domain and y = f(x) is an element of the Range. The function in the previous example can be written as ordered pairs (2, 12), (4, 14), (6, 16), (8, 18).

The graph of this function is a straight line:

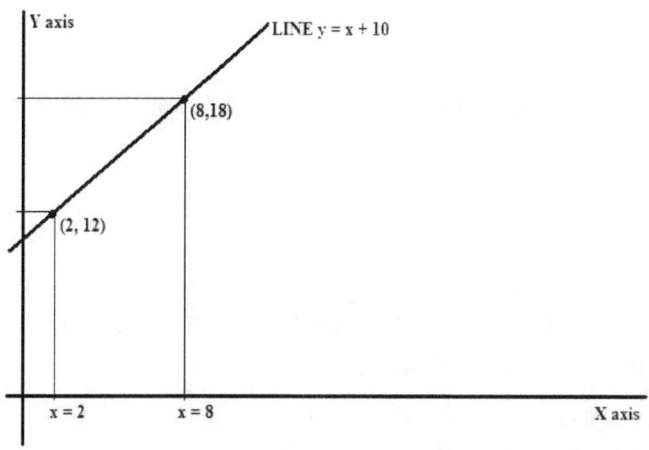

Example:

The function f(x) on the Domain {1, 2, 3, 4, 5} is defined by the formula f(x) = 2x − 5. For each x in the domain there is exactly one y = 2x − 5 in the range:

f(1) = 2(1) − 5 = − 3, point (1, − 3)
f(2) = 2(2) − 5 = −1, point (2, − 1)
f(3) = 2(3) − 5 = 1, point (3, 1)
f(4) = 2(4) − 5 = 3, point (4, 3)
f(5) = 2(5) − 5 = 5, point (5, 5)

The Range of this function is set R = {− 3, − 1, 1, 3, 5}.The graph of this function is a line through points (1, − 3), (2, − 1), (3, 1), (4, 3), (5, 5):

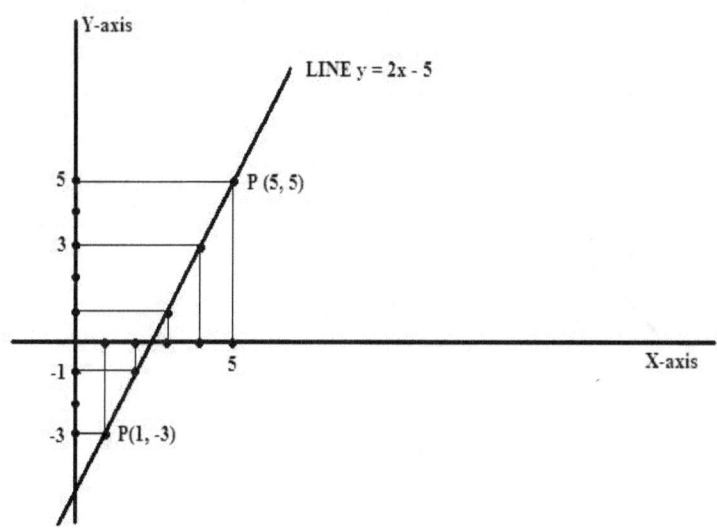

Any function f(x) can be represented as a graph on a coordinate plane consisting of ordered pairs (x, y) where x is in the Domain and y is in the Range. The coordinate y is the value of the function f(x). That is, y = f(x).

It is important to see function as a relation between a set of points and a graph. That is, the set of ordered pairs (points) on a plane.

Intercepts:

The y-intercept is the point where the graph of function "cuts through" the y-axis. The x-intercept is the point where the graph of function "cuts through" x–axis. Function may have more than one x-intercept. The x-intercepts are called "roots" or "zeroes" of function because at each x-intercept the value of y is zero:

x-intercept is a point (x, 0), y-intercept is a point (0, y)

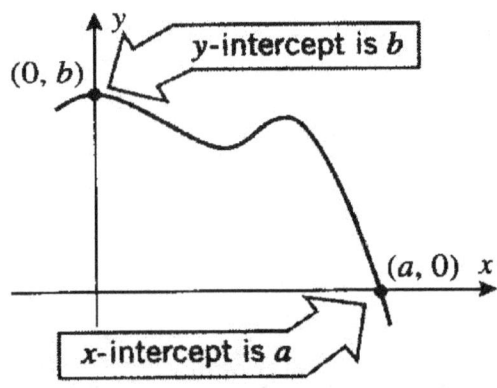

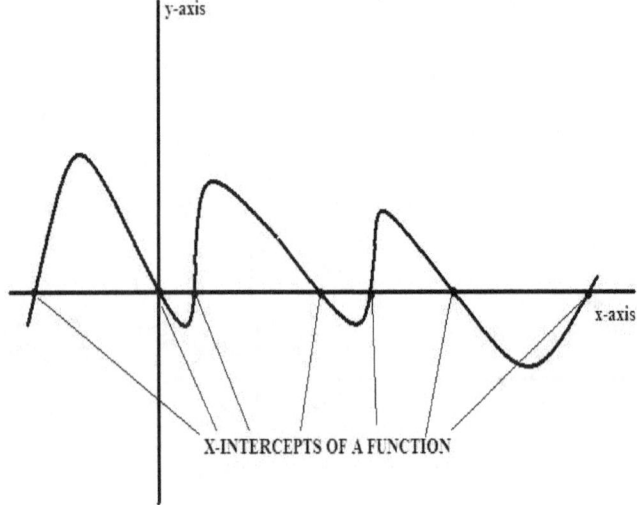

VERTICAL LINE TEST: IS IT FUNCTION?

Recall that in function every x must be assigned to only one y. Thus, for each x there is only one point (x, y) on a graph. Therefore, any vertical line would intersect the graph of function at one point only. If some vertical line intersects the graph of some relation in more than one point then this relation is not function.

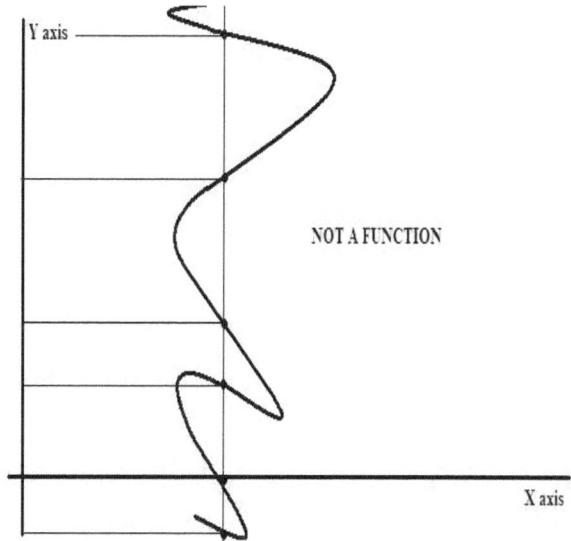

A relation between two sets is function only if each element of the Domain is assigned to exactly one element of the Range.

HORIZONTAL LINE TEST: IS IT A ONE–TO–ONE FUNCTION?

When each element of the Domain is assigned to exactly one, unique element of the Range, and each element of the Range has exactly one, unique element of the Domain assigned to it, such function is called one-to-one function. Any vertical line and any horizontal line can intersect the graph of such function at only one point.

If any horizontal line intersects the graph of f(x) at exactly one point then this graph is a graph of a one-to-one function. If there is a horizontal line that intersects the graph in more than one point then this function is not one-to-one. Because if there is a horizontal line that "cuts" the graph of function f(x) at more than one point then there are more than one x for each value of y. That is, more than one x is assigned to the same y. Such relation may be function, but not one-to-one function.

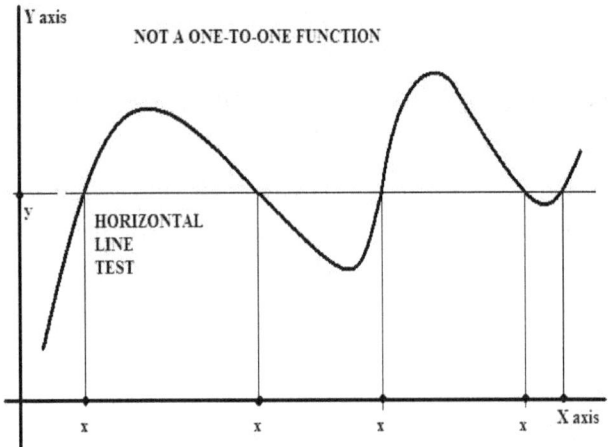

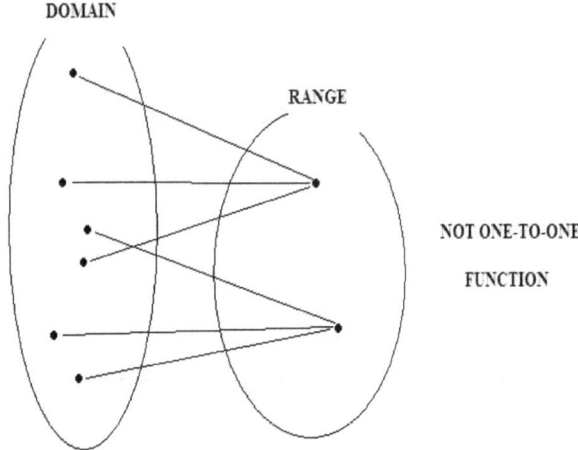

Definition: function f(x) is said to be one-to-one (*injective*) if and only if f(a) = f(b) implies that a = b, where a and b are two elements in the domain of the function.

Example:

Determine if function f(x) = 2x + 3 where x is a real number is one-to-one function.

If x is a real number then f(x) = 2x + 3 is also a real number. Thus, the domain of f(x) is the set of all real numbers and the range is the set of all real numbers.

By the definition, function f(x) would be one-to-one if f(a) = f(b) implies that a = b. Suppose that f(a) = f(b). Then 2a + 3 = 2b + 3. By subtracting 3 from both sides, we obtain 2a = 2b. By dividing both sides by 2, we obtain a = b.

Thus, the assumption that f(a) = f(b) implies that a = b. Therefore, f(x) = 2x + 3 is one-to-one function.

Example:

Determine if $f(x) = x^2$ is one-to-one function for all real x.

Suppose that $f(a) = f(b)$. Then $a^2 = b^2$. It seems that by taking the square root of both sides, we could obtain $a = b$, but this is not so because square root of a number may be positive or negative.

For example, $f(1) = 1^2 = 1$ and $f(-1) = (-1)^2 = 1$, but -1 is not equal to 1. Here, the assumption that $f(a) = f(b)$ does not imply that $a = b$, so function $f(x) = x^2$ is not one-to-one.

ARITHMETIC COMBINATIONS AND COMPOSITIONS OF FUNCTIONS:

Just as two numbers can be combined by operations of addition, subtraction, multiplication and division to form other numbers, two functions can be combined to create new functions. For example, $f(x) = 2x - 3$ and $g(x) = x^2$ can be combined to form the sum, difference, product, and quotient of $f(x)$ and $g(x)$:

$f(x) + g(x) = (2x - 3) + (x^2) = x^2 + 2x - 3$
$f(x) - g(x) = (2x - 3) - (x^2) = -x^2 + 2x - 3$
$f(x)g(x) = (2x - 3)(x^2) = 2x^3 - 3x^2$
$f(x)/g(x) = (2x - 3)/(x^2) = 2x/x^2 - 3/x^2 = 2/x - 3/x^2, x \neq 0$

The domain of each new function will consist of all the numbers that the domains of $f(x)$ and $g(x)$ have in common (*i.e.*, the intersection of the domains of $f(x)$ and $g(x)$). Another way of combining two functions is to form a composition of one with the other. The composition of $f(x)$ and $g(x)$ is written $f(g(x))$, the composition of $g(x)$ and $f(x)$ is written $g(f(x))$.

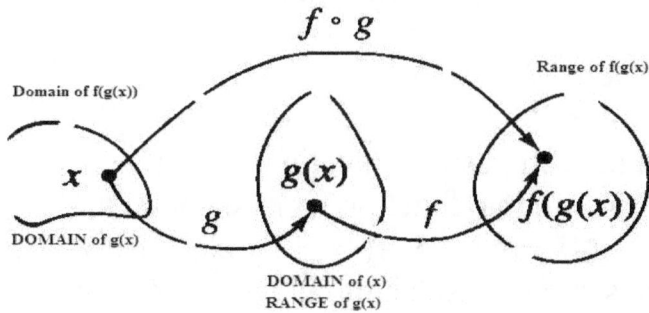

The domain of the composition function $f(g(x))$ is the set of all x such that $g(x)$ is in the domain of $f(x)$.

Example:

Let $f(x) = 2x + 1$ and $g(x) = 5x$. Find the composition function $f(g(x))$.

Composition function $f(g(x))$ is simply the function $f(x)$ whose domain is the range of the function $g(x)$. Substituting $g(x)$ instead of x in the original function $f(x)$ we obtain composition function $f(g(x))$: $f(g(x)) = f(5x) = 2(5x) + 1 = 10x + 1$.

Example:

Let $f(x) = 2x$ be defined on the domain $\{2, 4, 6, 8\}$. Let $g(x) = x + 4$.
We must substitute $g(x) = x + 4$ instead of x in $f(x) = 2x$, so that composition function $f(g(x)) = f(x + 4) = 2(x + 4) = 2x + 8$.

Example:

Let $m(x) = x^5$ and $w(x) = x - 4$. Find the composition function $m(w(x))$.
Substituting $w(x)$ instead of x in $m(x)$ we obtain $m(w(x)) = [w(x)]^5 = (x - 4)^5$.

Example:

Let $f(x) = 3x - 2$ and $g(x) = x^2 + 1$. Find the composition function $f(g(x))$.
Substituting $g(x)$ instead of x in $f(x) = x^2 + 1$ we obtain composition function
$f(g(x)) = 3(x^2 + 1) - 2 = 3x^2 + 3 - 2 = 3x^2 + 1$.

Example:

Let $f(x) = x^2$ and $g(x) = 3x + 4$. Find the composition function $f(g(x))$.
Substituting $g(x)$ into the original function $f(x)$ we obtain the composition function
$f(g(x)) = f(3x + 4) = (3x + 4)^2 = 9x^2 + 12x + 12x + 16 = 9x^2 + 24x + 16$.

Example:

Let $f(x) = x^2 + 5$ and $g(x) = 4x + 1$. Find the composition functions $f(g(x))$ and $g(f(x))$. Substituting $g(x)$ instead of x in the original function $f(x)$ we obtain the composition function $f(g(x)) = f(4x + 1) = (4x + 1)^2 + 5 = (4x + 1)(4x + 1) + 5 = 16x^2 + 8x + 1 + 5 = 16x^2 + 8x + 6$. In the same way, substituting $f(x)$ instead of x in the original function $g(x)$ we obtain the composition function
$g(f(x)) = g(x^2 + 5) = 4(x^2 + 5) + 1 = 4x^2 + 21$.

Example:

Let $f(x) = (3x + 1)^2$ and $g(x) = x - 2$. Find the composition functions $f(g(x))$ and $g(f(x))$.

Substituting $g(x)$ instead of x in the original function $f(x)$ we obtain:
$f(g(x)) = f(x - 2) = [3(x - 2) + 1]^2 = [3x - 6 + 1]^2 = [3x - 5]^2 = 9x^2 - 30x + 25$.

Substituting $f(x)$ instead of x in the original function $g(x)$ we obtain:
$g(f(x)) = g(3x + 1)^2 = (3x + 1)^2 - 2 = (9x^2 + 6x + 1) - 2 = 9x^2 + 6x - 1$.

Example:

Let $f(x) = 3x + 1$ and $g(x) = \sqrt{x}$, find the composition functions $f(g(x))$ and $g(f(x))$.

$f(g(x)) = f(g(x)) = f(\sqrt{x}) = 3\sqrt{x} + 1$,
$g(f(x)) = g(f(x)) = g(3x + 1) = \sqrt{(3x + 1)}$.

Note that $f(g(x))$ is not equal to $g(f(x))$.

Exercises:

Find the composition functions $f(g(x))$ and $g(f(x))$ for the following functions:

1. $f(x) = x + 5$, $g(x) = 2x$

2. $f(x) = 2x - 1$, $g(x) = x^2$

3. $f(x) = x - 3$, $g(x) = x^{1/2}$ [$x^{1/2} - 3$, $(x - 3)^{1/2}$]

4. $f(x) = x/2$, $g(x) = 4x + 10$

5. $f(x) = 2x$, $g(x) = x^4$, find $f(g(4))$ and $g(f(4))$ [512, 4096]

6. $f(x) = 2x + 3$, $g(x) = x - 3$

7. $f(x) = x + 3$, $g(x) = x^2 - 3x + 2$, find $f(g(2))$ [3]

INVERSE FUNCTIONS:

If the functions f(x) and g(x) are both one-to-one functions and the composition f(g(x)) = g(f(x)) = x, then function f(x) is the inverse of g(x) and function g(x) is the inverse of f(x). That is, f and g are inverses of each other.

It means that the function f "goes" from the domain A to the range B, whereas the function g "goes" from B to A. The function g would be the "opposite" of the function f in the sense that what was the domain of f would become the range of g and what was the range of function f would become the domain of function g:

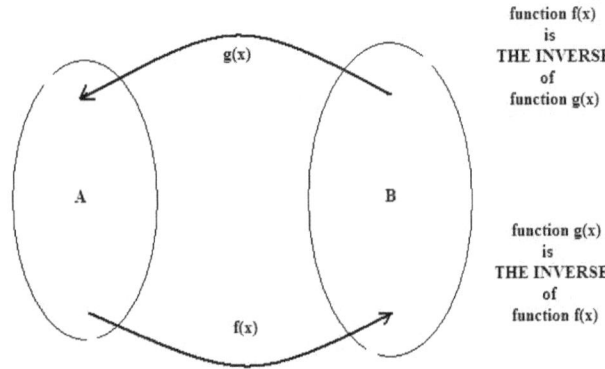

Recall that function can be represented by a set of ordered pairs (x, y). For example, function f(x) = x + 4 from the set A = {1, 2, 3, 4} to the set B = {5, 6, 7, 8} can be written {(1, 5), (2, 6), (3, 7), (4, 8)}. By interchanging x and y coordinates we obtain the inverse function of f, which is denoted f^{-1}.

The inverse function of f(x) in this example is function from the set B to the set A, and can be written {(5, 1), (6, 2), (7, 3), (8, 4)}.

Example:

Let f(x) = {(0, 3), (1, 5), (3, 9), (5, 13)}, find the inverse of f(x).

The inverse reverses the ordered pairs: $f^{-1}(x)$ = {(3, 0), (5, 1), (9, 3), (13, 5)}.

The inverse functions are simply "undoing" each other, so the compositions $f(f^{-1}(x))$ and $f^{-1}(f(x))$ are both equal to x:

$f(f^{-1}(x)) = f(x-4) = (x-4) + 4 = x$ and $f^{-1}(f(x)) = f^{-1}(x+4) = (x+4) - 4 = x$

Function f(x) has the inverse $f^{-1}(x)$ if and only if each x is assigned to exactly one y = f(x). Then, the inverse function $f^{-1}(x)$ will assign y = f(x) back to x. Thus, f(x) is one-to-one function. Only one-to-one functions have inverses.

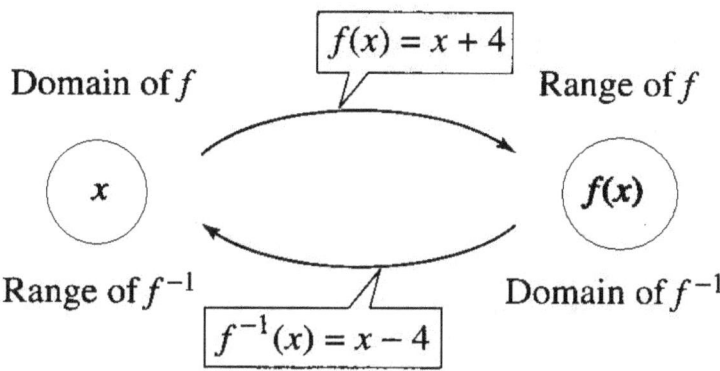

Finding Inverse Functions Algebraically:

To find the inverse of function algebraically simply follow these steps:

1. Replace f(x) by y
2. Interchange x and y
3. Solve for y

Example:

Find the inverse of f(x) = 2x + 5.

1. Replacing f(x) with y we obtain y = 2x + 5
2. Interchanging x and y we obtain x = 2y + 5
3. Solving for y we obtain x − 5 = 2y and y = (x − 5)/2.

Example:

Find the inverse of f(x) = 4x − 5.

1. Replacing f(x) with y we obtain y = 4x − 5
2. Interchanging x and y we obtain x = 4y − 5
3. Solving for y we obtain 4y = x + 5 and y = (x + 5)/4.

Therefore, the inverse of f(x) = 4x − 5 is $f^{-1}(x)$ = (x + 5)/4.

Example:

Find the inverse of $f(x) = 3(x + 4)$.

1. Replacing $f(x)$ with y we obtain $y = 3(x + 4)$
2. Interchanging x and y we obtain $x = 3(y + 4)$
3. Solving for y we obtain $x/3 = y + 4$ and $y = (1/3)x - 4$

Example:

Consider the function $h(x) = x^2$. Does it have the inverse?

Interchanging x and y, we obtain $x = y^2$. Solving for y, we obtain two solutions: positive and negative square roots of x. That is, $y = +\sqrt{x}$ or $y = -\sqrt{x}$. Since for every x-value there are two different y-values, this is not function. Therefore, $h(x) = x^2$ does not have the inverse. Function may have at most one inverse, so if unction does have the inverse, it is unique.

Example:

Verify that $f(x) = (1/2)x - 2$ is the inverse of $f(x) = 2x + 4$ without finding the inverses. Functions $f(x)$ and $g(x)$ are inverses of each other if $f(g(x)) = x = g(f(x))$.

$f(g(x)) = (1/2)(2x + 4) - 2 = x + 2 - 2 = x$
$g(f(x)) = 2((1/2)x - 2) + 4 = x - 4 + 4 = x$
Therefore, $f(x)$ and $g(x)$ are inverses of each other.

Exercises:

Find the inverse function $f^{-1}(x)$, if it exists, for the following functions:

1. $f(x) = 3x$

2. $f(x) = 4x + 1$

3. $f(x) = 2x + 5$ [$(x - 5)/2$]

4. $f(x) = 5 + 2x$

5. $f(x) = 4(x + 1$

FAMILIES OF FUNCTIONS:

Functions are divided into several types:

1. Constant functions $f(x) = c$, where c is a real number
2. Linear function $f(x) = mx + b$, which is an equation of a line
3. Power (exponential) functions $f(x) = x^n$
4. Polynomial function $f(x)$, which is any polynomial
5. Rational functions $f(x) = P/Q$, P and Q are polynomials, $Q \neq 0$
6. Trigonometric functions

1. Constant Function:

Function whose values are always the same is called a "constant function". All numbers x in the Domain are assigned to only one number y in the Range. That is, the range of the function has only one value c. The graph of the constant function is a straight horizontal line $f(x) = c$ (c is a real number).

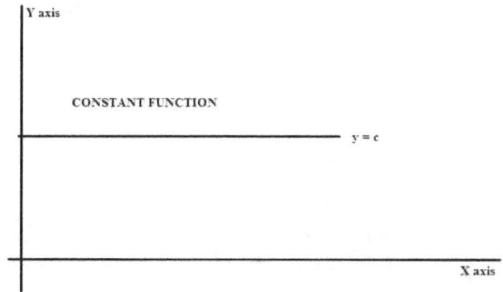

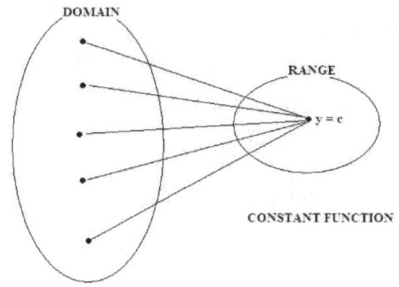

2. Linear Function:

Any straight line whatsoever is a graph of some linear function $f(x)$.
The linear function $f(x)$ is simply a linear equation, such as

$y = 4x$, $y = x + 5$, $y = 3x - 7$, $y = 2(x + 1)$, $y = 6(3x + 5) - 7x$, and so on.

An equation of a line (linear function) is written $y = f(x) = mx + b$ where m is the slope of the line and b is the y-intercept of this line. The slope m of the line is the distance "up" divided by the distance "across". Equivalently, the slope m of the line between two points (x_1, x_2) and (y_1, y_2) is often called "rise over run", which is $m = (y_2 - y_1)/(x_2 - x_1)$.

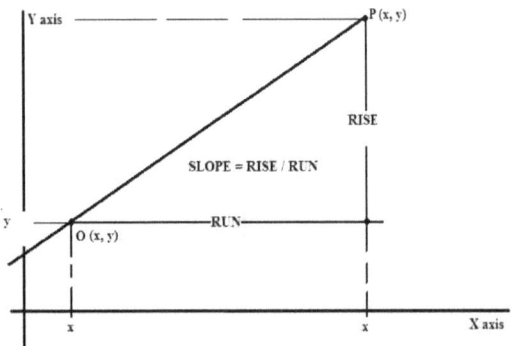

3. Power Function:

Power (or exponential) functions are of the form $f(x) = x^n$ where n is a real number. The graph of such function is a curve. For example:

Degree n = 2   Degree n = 3   Degree n = 4   Degree n = 5

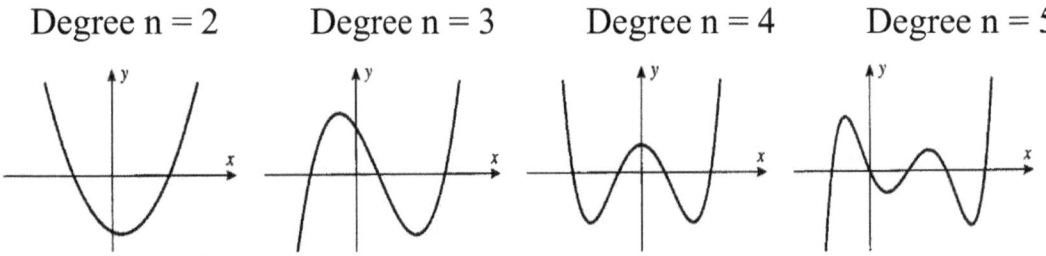

4. Polynomial Function:

A polynomial function is expressible as a sum of several terms cx^n where c and n are real numbers, n is 0 or positive. For example, $2x^4 + 3x^3 - 8x^2 + 5x - 7$ is a polynomial function where constants are 2, 3, − 8, 5, − 7 and exponents n are 4, 3, 2, 1, 0. The graph of such function is a curve.

5. Rational Function:

Rational function is a ratio of two polynomials P(x) and Q(x). That is, $f(x) = P(x)/Q(x)$, where the polynomial Q may not be equal to zero (because division by zero is not defined). The graph of such function is a curve or several curves.

6. Basic Trigonometric functions are sin x, cos x, tan x, cot x, sec x, csc x.

SYMMETRY, EVEN AND ODD FUNCTIONS:

Function f(x) is called an even function if exchanging $(-x)$ for x does not change the function. That is, $f(-x) = f(x)$ for all x in the domain of the function. Function f(x) is called an odd function if exchanging $(-x)$ for x changes the sign of the function. That is, $f(-x) = -f(x)$ for all x in the domain of the function.

Examples:

Function $f(x) = x^2$ is even because $f(-x) = (-x)^2$ which is the same as $f(x) = x^2$.
Function $f(x) = x^3$ is odd because $f(-x) = (-x)^3$ which is equal to $-f(x) = -x^3$.

Saying that function f(x) is even is equivalent to saying that its graph is symmetric about y-axis because for every point (x, y) on the graph of the function there is a corresponding point $(-x, y)$. The y-coordinate does not change, only the x-coordinate changes to the negative x. Saying that function f(x) is odd is equivalent to saying that its graph is symmetric with respect to the origin (0, 0) because for every point (x, y) on the graph there is a corresponding point $(-x, -y)$.

Example:

The function $f(x) = x^2 - 1/x^2 + 5$, $x \neq 0$, is symmetric about y-axis (even) because substituting $(-x)$ instead of x does not change the function:
$f(-x) = (-x)^2 - 1/(-x)^2 + 5 = x^2 - 1/x^2 + 5 = f(x)$.

Example:

The function $f(x) = x + 1/x$, $x \neq 0$, is symmetric with respect to the origin (odd) because substituting $(-x)$ instead of x changes the sign of the function:
$f(-x) = (-x) + 1/(-x) = -x - 1/x = -(x + 1/x) = -f(x)$.

Example:

The function $f(x) = (x + 1)/x$ is neither even nor odd:
$f(-x) = ((-x) + 1)/(-x) = 1 - 1/x = (x - 1)/x$, which is not equal to f(x) and is not equal to $-f(x)$, either.

ZEROES OF FUNCTION:

The values of x such that $f(x) = 0$ are called "zeroes" of the function $f(x)$. If zeros exist, they are simply x-intercepts of the function.

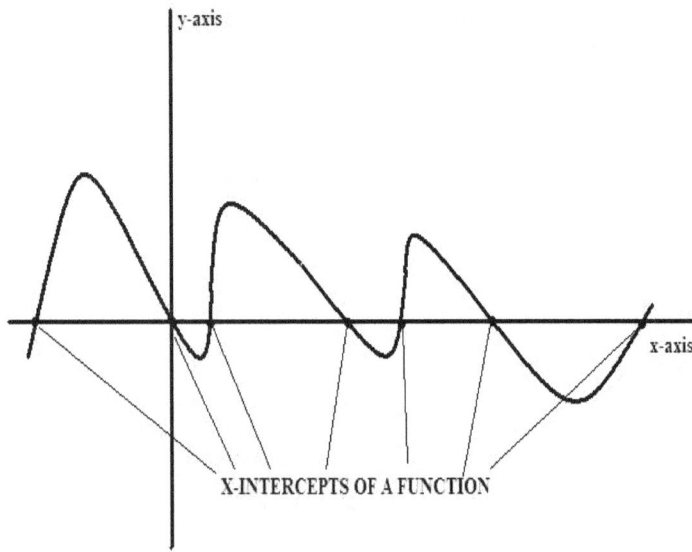

Example:

Find zeroes of $f(x) = (x - 1)(x + 1)(2x - 3)(5x + 4)$.
To find zeros of function, we need to find the values of x that make $f(x) = 0$.
Let $(x - 1)(x + 1)(2x - 3)(5x + 4) = 0$. Then, $x - 1 = 0$ or $x + 1 = 0$ or $2x - 3 = 0$ or $5x + 4 = 0$. Thus, $x = 1$ or $x = -1$ or $x = 3/2$ or $x = -4/5$. These numbers, which are x-intercepts of function $f(x)$, are called "zeroes" or "roots" of $f(x)$.

Example:

Find zeros of $f(x) = x^2 + x - 6$.

To find zeros of function, we need to find which values of x make $f(x) = 0$.
By factoring $x^2 + x - 6$, we obtain $(x + 3)(x - 2)$.
If $(x + 3)(x - 2) = 0$, then $x = -3$ or $x = 2$.

These are the "zeroes" of function $f(x) = x^2 + x - 6$.

CHAPTER 2 – LIMITS

THE CONCEPT OF A LIMIT:

A very important, basic concept in Calculus is that of the limit of function: if f(x) is function of x, what happens to the value of the function y = f(x) as x approaches some number?

For example, if $f(x) = x^2 + 3$, what is the value of y = f(x) as x approaches 5?

x	y
1	4
3	11
4	19
4.5	23.25
4.7	25.09
4.9	27.01
4.96	27.6016
4.98	27.8004
4.99	27.9001

As x approaches 5, the value of y = f(x) approaches $5^2 + 3 = 28$.
Number 28 is called "the limit of $f(x) = x^2 + 3$ as x approaches 5".
The notation is $\lim_{x \to 5} (x^2 + 3) = 28$.

Another example of a limit irrational number *e* (the base of a natural logarithm *ln*). The value of *e* can be approximated by the formula $(1 + 1/n)^n$ where n is a real number:

n	$(1 + 1/n)^n$
5	$(1 + 1/5)^5 = 2.48832$
10	$(1 + 1/10)^{10} = 2.59374246$
100	$(1 + 1/100)^{100} = 2.704813829$
1000	$(1 + 1/1000)^{1000} = 2.716923932$
10000	$(1 + 1/10000)^{10000} = 2.718145927$

As n increases indefinitely, the value of $(1 + 1/n)^n$ gives a better and better approximation for *e*, which is 2.718281828... . That is, *e* is the limit of $(1 + 1/n)^n$ as n approaches infinity: $\lim_{n \to \infty} (1 + 1/n)^n = e$.

Example:

Let $f(x) = x^2 - x + 1$, then $f(2) = 2^2 - 2 + 1 = 4 - 2 + 1 = 3$.
As the values of x approach 2, the values of y approach $f(2) = 3$.

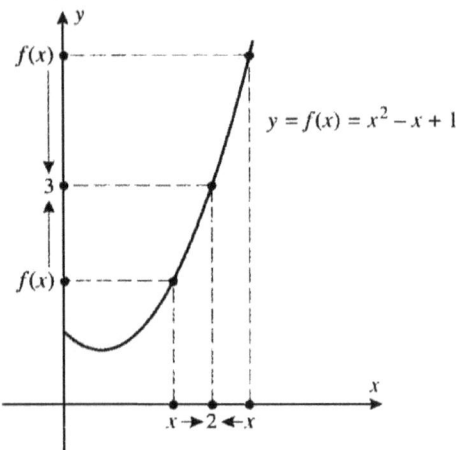

We can approach the limiting value of x from the left as well as from the right. Such limits are called *one–sided limits*.

Limit of f(x) as x approaches *a* from the right: $\lim_{x \to a+} f(x) = L$
Limit of f(x) as x approaches *a* from the left: $\lim_{x \to a-} f(x) = L$

The two–sided limit $\lim_{x \to a} f(x) = L$ exist at a point (*a*, f(*a*)) if and only if *both* one–sided limits exist at that point and have the same value. Intuitively, it is obvious that if we obtain different values for a limit by approaching the same point from the left or the right, then the limit does not exist. For some functions, the limit of f(x) as x approaches some number *a* may exist but the function itself may be undefined at that point. For example, $f(x) = (x^2 - 1)/(x - 1)$ is defined for all x except x = 1. This function is undefined for x = 1, but the limit of the function as x approaches 1 exists and is equal to 2:

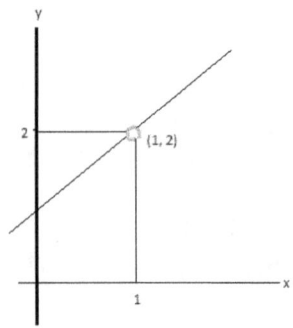

To evaluate this limit algebraically, we need to factor the numerator and cancel the common term, thus obtaining:

$\lim_{x \to 1} (x^2 - 1)/(x - 1) = \lim_{x \to 1} (x - 1)(x + 1)/(x - 1) = \lim_{x \to 1} (x + 1) = 1 + 1 = 2.$

Example:

The *greatest integer* function [x] is defined as the largest integer that is less than or equal to x, that is [x] = n, where n ≤ x.

For example, [1.6] = 1, [2.37] = 2, [5] = 5, [– 5] = – 5, [– 3.8] = – 4.

The limit of the greatest integer function does not exist because one–sided limits are not the same: limit of [x] as x approaches 2 from the right is 2, $\lim_{x \to 2+}[x] = 2$. However, limit of [x] as x approaches 2 from the left is 1, $\lim_{x \to 2-} [x] = 1.$

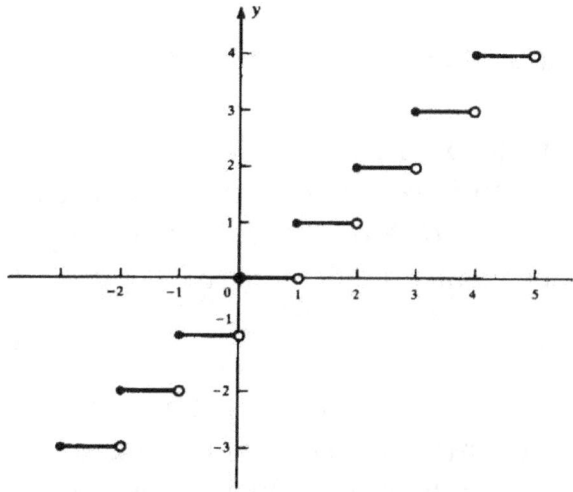

The greatest integer function is an example of step function. Step functions have points of discontinuity (holes or breaks in the graph of the function) where they jump from one value to another without taking on any intermediate values.

Properties of Limits:

Let f(x) and g(x) be any two functions such that limit of f(x) as x approaches some number c is equal to L and limit of g(x) as x approaches some number c is equal to M. That is, $\lim_{x \to c} f(x) = L$ and $\lim_{x \to c} g(x) = M$, where L and M are real numbers (that is to say, both limits exist). Then the properties of limits can be defined by these rules:

1. $\lim_{k \to c} k = k$, for any constant k.
2. $\lim_{x \to c} x = c$.
3. $\lim_{x \to c} [f(x) + g(x)] = \lim_{x \to c} f(x) + \lim_{x \to c} g(x) = L + M$
4. $\lim_{x \to c} [f(x) - g(x)] = \lim_{x \to c} f(x) - \lim_{x \to c} g(x) = L - M$
5. $\lim_{x \to c} kf(x) = k \cdot \lim_{x \to c} f(x) = kL$, for any constant k.
6. $\lim_{x \to c} [f(x) \cdot g(x)] = \lim_{x \to c} f(x) \cdot \lim_{x \to c} g(x) = LM$
7. $\lim_{x \to c} [f(x)/g(x)] = \lim_{x \to c} f(x) \div \lim_{x \to c} g(x) = L/M$
8. $\lim_{x \to c} [f(x)]^{1/n} = [\lim_{x \to c} f(x)]^{1/n} = [L]^{1/n}$, where $L > 0$ for all even n.
9. $\lim_{x \to c} f(x) = f(c)$ for all polynomial functions f(x).
10. $\lim_{x \to c} r(x) = r(c)$ for all rational functions r(x) with denominator $\neq 0$ at x = c.

Examples:

$\lim_{x \to 2} (x^2 - 5x - 1) = 2^2 - 5(2) - 1 = -7$

$\lim_{x \to 4} 2x/(3x + 1) = 2(4)/(3(4) + 1) = 8/13$

$\lim_{x \to -1} (2x^2 + 3)^{1/2} = (\lim_{x \to -1} (2x^2 + 3))^{1/2} = (2(-1)^2 + 3)^{1/2} = (5)^{1/2}$

When the limit of a rational function $\lim_{x \to c} f(x)/g(x)$ has a form 0/0 as x approaches some number c, such limit is called an *indeterminate form*. Function must be re-written so that denominator is not equal to zero. This can be done by factoring and canceling common term and then evaluating the limit of the polynomial.

If both numerator and denominator are zero when x = c, it means that both numerator and denominator have a common factor (x − c). Then we can factor and cancel this common factor, thus obtaining simple polynomial which is easy to evaluate.

Example:

Evaluate $\lim_{x \to 2} (x^2 - 4)/(x - 2)$.

Both numerator and denominator are zero when x = 2.
By factoring $(x^2 - 4)$, we obtain $(x - 2)(x + 2)$.

Therefore, $\lim_{x \to 2} (x^2 - 4)/(x - 2) = \lim_{x \to 2} (x - 2)(x + 2)/(x - 2)$.
By cancelling common factor (x − 2), we obtain $\lim_{x \to 2} (x + 2) = 2 + 2 = 4$.

Since function $(x^2 - 4)/(x - 2)$ is discontinuous at $x = 2$, there is a gap in the graph of the function at the point corresponding to $x = 2$. *Note:* it is important to understand that the limit of the function as x approaches 2 exists and is equal to 4. But the function is not defined at the point corresponding to $x = 2$.

Example:

Evaluate $\lim_{x \to -3} (x^2 + x - 6)/(x + 3)$.
Both numerator and denominator are zero when $x = -3$
By factoring numerator $(x^2 + x - 6)$, we obtain $(x + 3)(x - 2)$
Therefore, $\lim_{x \to -3} (x + 3)(x - 2)/(x + 3) = \lim_{x \to -3} (x - 2) = -3 - 2 = -5$.

Since the function $(x^2 + x - 6)/(x + 3)$ is discontinuous at $x = -3$, there is a gap in the graph of the function at that point. *Note:* the limit of the function as x approaches -3 exists and is equal to -5. The limit at $x = -3$ exists, but the function is not defined at the point corresponding to $x = -3$.

Example:

Evaluate $\lim_{x \to 4} (x^2 - 5x + 4)/(x^2 - 2x - 8)$.

Both numerator and denominator are equal to zero when $x = 4$. By factoring both numerator and denominator, we obtain $(x - 4)(x - 1)/(x + 2)(x - 4)$, but function $f(x) = (x^2 - 5x + 4)/(x^2 - 2x - 8)$ is not defined at $x = 4$ and $x = -2$ and therefore $x \neq 4$ and $x \neq -2$. Thus,

$\lim_{x \to 4} (x^2 - 5x + 4)/(x^2 - 2x - 8) = \lim_{x \to 4} (x - 4)(x - 1)/(x + 2)(x - 4)$

By cancelling common term $(x - 4)$, we obtain $\lim_{x \to 4} (x - 1)/(x + 2) = 3/6 = 1/2$.

Note: the limit of the function as x approaches 4 exists and is equal to 1/2. But the function is not defined at the points corresponding to $x = -2$ and $x = 4$. That is, the limit of the function as x approaches 4 is equals to 1/2, but the function itself is not defined at $x = 4$.

Suppose that we consider a limit as x approaches some different number, say, -3. Then $\lim_{x \to -3} (x^2 - 5x + 4)/(x^2 - 2x - 8) = \lim_{x \to -3} (x - 1)/(x + 2) = (-4)(-1) = 4$. Function is defined at the limiting value $x = -3$ and the value of function $f(x)$ is 4.

CONTINUITY:

Function f(x) may be discontinuous at a point x = c in three basic ways:

1. Function is undefined at c
2. Limit of f(x) does not exist as x approaches c
3. Value of the function at c is different from the limit at c

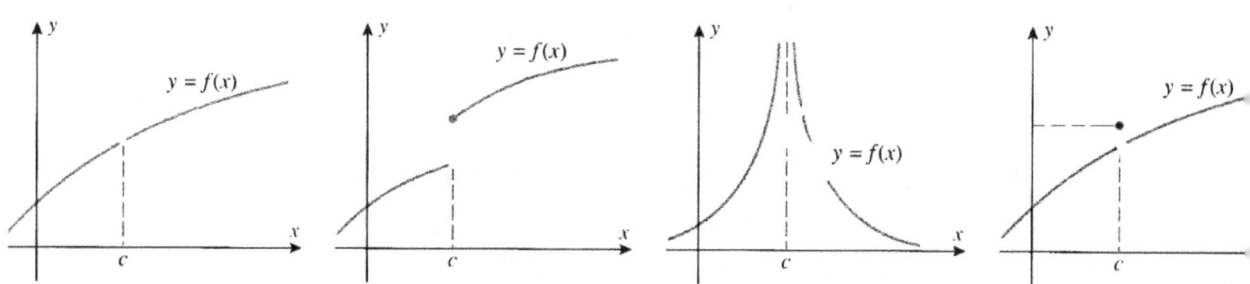

Function f(x) is said to be continuous if the graph of f(x) has no holes or breaks in it. Formally stated, Function f(x) is continuous at the point c if the following conditions are satisfied:

1. f(c) is defined (that is, f(c) exists)
2. $\lim_{x \to c} f(x)$ exists
3. $\lim_{x \to c} f(x) = f(c)$

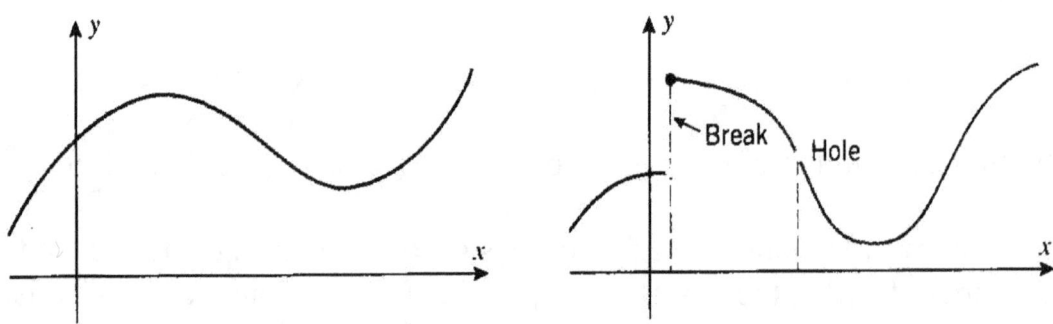

All polynomial functions are continuous everywhere. Rational functions of the form P(x)/Q(x), where P(x) and Q(x) are polynomials and Q(x) ≠ 0, are continuous everywhere. Functions of the form P(x)/Q(x) are discontinuous at the point(s) where the denominator Q(x) equals to zero.

Example:

For which values of x is function $f(x) = (5x + 15)/(x - 3)(x - 2)$ discontinuous and at which intervals it is continuous?

This function is undefined (discontinuous) when the denominator $(x - 3)(x - 2)$ is equal to zero. The denominator $(x - 3)(x - 2)$ is zero when $x - 3 = 0$ or $x - 2 = 0$. Therefore, $f(x)$ is discontinuous at $x = 3$ and $x = 2$.

This function is continuous on the intervals from negative infinity to 2, from 2 to 3, and from 3 to positive infinity: $(-\infty, 2)$, $(2, 3)$, and $(3, +\infty)$.

Example:

For which values of x is $f(x) = (2x^4 + 5x^3 - 7x^2 - 5x + 1)/(x^2 - 9)$ discontinuous and at which intervals it is continuous?

The denominator $x^2 - 9$ is equal to zero when $x^2 = 9$. Therefore, $x = \sqrt{9}$ and so function $f(x)$ is discontinuous at $x = 3$ and $x = -3$.

This function is continuous on the intervals from negative infinity to -3, from -3 to 3, and from 3 to positive infinity: $(-\infty, -3)$, $(-3, 3)$, and $(3, +\infty)$.

Example:

Find the intervals on which function $f(x) = x/(x^2 - x - 2)$ is continuous.

This function is not continuous when the denominator $x^2 - x - 2$ is equal to zero. Factoring $x^2 - x - 2$ we obtain $(x + 1)(x - 2) = 0$. Function $f(x) = x/(x^2 - x - 2)$ is not continuous at two points: $x = -1$ and $x = 2$.

Function is continuous on the intervals from negative infinity to -1, from -1 to 2, and from 2 to positive infinity: $(-\infty, -1)$, $(-1, 2)$, $(2, +\infty)$.

Continuity on a Closed Interval:

Function $f(x)$ is continuous on closed interval $[a, b]$ if it is continuous on open interval (a, b) and the limit of $f(x)$ as x approaches a from the right and the limit of $f(x)$ as x approaches b from the left both exist:
$\lim_{x \to a+} f(x) = f(a)$ and $\lim_{x \to b-} f(x) = f(b)$

Note: If function is not continuous at a point where x = c, it is certainly not differentiable at this point. But it is possible for the function to be continuous at x = c and *not* be differentiable at this point. (For example, when tangent line at point x = c is vertical).

Removable and Nonremovable Discontinuities:

A discontinuity of function f(x) at x = c is called *removable* if the limit of f(x) as x approaches c exists, i.e., $\lim_{x \to c} f(x) = k$, for some number k. A discontinuity of function f(x) at x = c is called *nonremovable* if the limit of f(x) as x approaches c does not exists, $\lim_{x \to c} f(x) = +\infty$ or $-\infty$.

Example:

Function f(x) = 5/x has a nonremovable discontinuity at x = 0 because f(x) is undefined at x = 0 and the limit of f(x) as x approaches 0 does not exist: $\lim_{x \to 0} 5/x = +\infty$.

Example:

Function $f(x) = (x - 6)/(x^2 - 36)$ has two discontinuities: x = 6 and x = -6. Discontinuity at x = -6 is nonremovable because $\lim_{x \to -6} (x - 6)/(x^2 - 36) = -\infty$. Discontinuity at x = 6 is removable because limit exists: $\lim_{x \to 6} (x - 6)/(x^2 - 36) = 1/(x + 6) = 1/12$.

Example:

Function $f(x) = (x + 2)/(x + 2)(x - 5)$ has a nonremovable discontinuity at x = 5 because $\lim_{x \to 5} (x + 2)/(x + 2)(x - 5) = \infty$. However, the discontinuity at x = -2 is removable because $\lim_{x \to -2} (x + 2)/(x + 2)(x - 5) = \lim_{x \to -2} 1/(x - 5) = -1/7$.

INFINITE LIMITS AND VERTICAL ASYMPTOTES:

If the values of f(x) increase indefinitely as x approaches number *a* either from the right or the left, then we write $\lim_{x \to a+} f(x) = +\infty$ or $\lim_{x \to a-} f(x) = +\infty$

If the values of f(x) decrease indefinitely as x approaches number *a* either from the right or the left, then we write $\lim_{x \to a+} f(x) = -\infty$ or $\lim_{x \to a-} f(x) = -\infty$

If both one sided limits increase/decrease indefinitely (without bound), then we write $\lim_{x \to a} f(x) = +\infty$ or $\lim_{x \to a} f(x) = -\infty$

Geometrically, if f(x) approaches infinity as x approaches some number *a* either from the right or from the left, then the graph of y = f(x) gets closer and closer to the vertical line x = *a* without ever crossing this line. This line is called *vertical asymptote*.

Examples of Vertical Asymptotes:

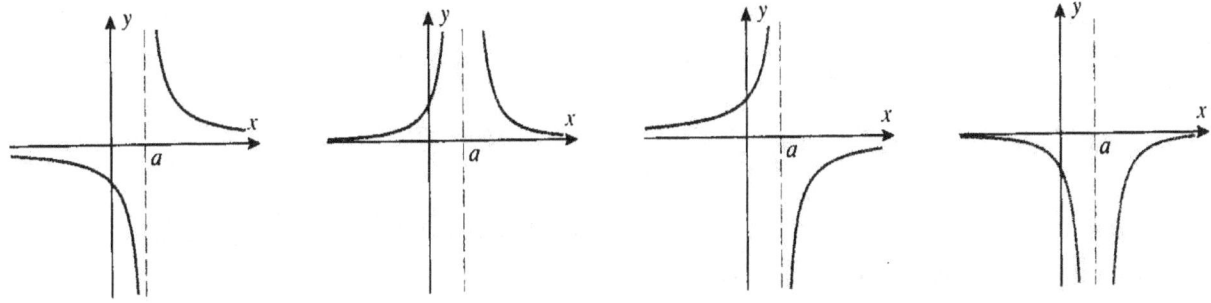

For a rational function R = f(x)/g(x), if the denominator g(c) = 0 but the numerator f(c) ≠ 0 when x = c, then line x = c is a vertical asymptote of the graph of function R(x).

Example:

Find points of discontinuity and identify vertical asymptotes of function $f(x) = (x^2 + x - 2)/(x^2 - 1)$.

Denominator $x^2 - 1 = (x - 1)(x + 1)$ thus f(x) is undefined for x = −1 and x = 1. Thus, f(x) has two discontinuities at x = −1 and x = 1.
If x = −1, numerator $x^2 + x - 2 = (-1)^2 + (-1) - 2 \neq 0$ but denominator is 0.
Thus, x = −1 is a vertical asymptote of the graph of $f(x) = (x^2 + x - 2)/(x^2 - 1)$.
If x = 1, numerator $x^2 + x - 2 = (1)^2 + (1) - 2 = 0$ and denominator is 0 (form 0/0).
Then $\lim_{x \to 1} (x^2 + x - 2)/(x^2 - 1) = \lim_{x \to 1} (x - 1)(x + 2)/(x - 1)(x + 1)$ which is equal to $\lim_{x \to 1} (x + 2)(x + 1) = 3/2$.

Since the limit of f(x) exists as x approaches 1, f(x) does not have an asymptote at x = 1. Function f(x) is simply discontinuous at x = 1. This discontinuity is removable because the limit at x =1 exist.

Example:

Find vertical asymptotes of $f(x) = (x^2 + 1)/(x^2 - 1)$.

By factoring the denominator, we obtain
$f(x) = (x^2 + 1)/(x^2 - 1) = (x^2 + 1)/(x - 1)(x + 1)$.

Since the denominator is zero for $x = -1$ and $x = 1$, $f(x)$ is not defined (discontinuous) at these points.

Numerator $(x^2 + 1)$ is not zero at either $x = -1$ or $x = 1$, $(-1)^2 + 1 = (1^2 + 1) = 2$. Therefore, $f(x)$ has vertical asymptotes at $x = -1$ and $x = 1$.

LIMITS AT INFINITY AND HORIZONTAL ASYMPTOTES:

If the values of f(x) get closer and closer to some number L as x increases indefinitely, then we write $\lim_{x \to +\infty} f(x) = L$

If the values of f(x) get closer and closer to some number L as x decreases indefinitely, then we write $\lim_{x \to -\infty} f(x) = L$

Geometrically, if $f(x) \to L$ as $x \to +\infty$, then the graph of f(x) gets closer and closer to a horizontal line $y = L$ without ever crossing it.

This line is called *horizontal asymptote*.

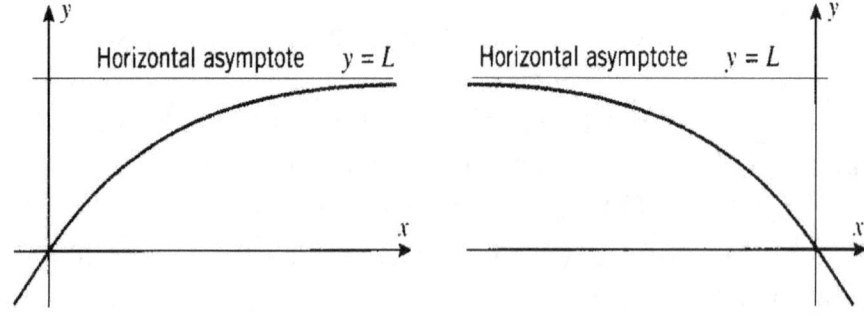

The limit of a polynomial function as x approaches infinity is the limit of its term of highest degree (leading term). This limit is infinity.

Example: $\lim_{x \to \infty} (5x^4 + 6x^3 - 8x^2 + x - 10) = \lim_{x \to \infty} (5x^4) = \infty$.

Given rational function f(x)/g(x), if the limits of both numerator and denominator approach infinity as x approaches infinity, then this is *indeterminate form* ∞/∞.
To evaluate $\lim_{x \to \infty} f(x)/g(x)$, we can divide both numerator and denominator by the highest power of x.

Example:

Evaluate $\lim_{x \to \infty} (2x - 1)/(x + 1)$.

Both $\lim_{x \to \infty} (2x - 1) = \infty$ and $\lim_{x \to \infty} (x + 1) = \infty$.
Therefore, by dividing both numerator and denominator by x, we obtain
$\lim_{x \to \infty} [(2x - 1)/x]/[(x + 1)/x] = \lim_{x \to \infty} (2x/x - (1/x))/(x/x + (1/x)) = 2/1 = 2$.

Note that (1/x) approaches zero as x approaches infinity: $\lim_{x \to \infty} (1/x) = 0$.

Line y = 2 is horizontal asymptote of the function (2x − 1)/(x + 1) to the right of zero. The limit of function as x approaches − ∞ is 2, $\lim_{x \to -\infty} (2x - 1)/(x + 1) = 2$.
Therefore, line y = 2 is also a horizontal asymptote to the left of zero as well.

The limit of a rational function as x approaches infinity may be equal to zero, a non-zero number, or the limit might not exist.

If the degree of the numerator is less than the degree of the denominator, then $\lim_{x \to \infty} f(x)/g(x) = 0$. Horizontal asymptote is the line y = 0 (x-axis).

If the degree of the numerator is equal to the degree of the denominator, then $\lim_{x \to \infty} f(x)/g(x) = k$.

Horizontal asymptote is the line y = a/b where a and b are leading coefficients of f(x) and g(x).

If the degree of the numerator is greater than the degree of the denominator, then $\lim_{x \to \infty} f(x)/g(x)$ does not exist. Function f(x)/g(x) has no horizontal asymptote.

Example:

Evaluate $\lim_{x \to \infty} (3x - 2x^2)/(7x^2 + 5)$.

The degree of the numerator and the denominator is the same, therefore the limit as x approaches infinity is a rational number. We can divide both numerator and denominator by the highest power of x, which is 2, to obtain

$\lim_{x \to \infty} (3x - 2x^2)/(7x^2 + 5) = \lim_{x \to \infty} (3/x - 2)/(7 + 5/x^2) = (0 - 2)/(7 + 0) = -2/7$.

Examples:

$\lim_{x \to \infty} x/(x^2 - 1) = \lim_{x \to \infty} (1/x)/(1 - (1/x^2)) = 0/1 = 0$.

$\lim_{x \to \infty} [(5/x) + (x/3)] = \lim_{x \to \infty} (5/x) + \lim_{x \to \infty} (x/3) = 0 + \infty = \infty$ (limit does not exist).

$\lim_{x \to \infty} (x^2 + 3)/(2x^2 - 1) = \lim_{x \to \infty} (1 + 3/x^2)/(2 - 1/x^2) = (1 + 0)/(2 - 0) = 1/2$.

$\lim_{x \to 2} (x^2 - 4)/(x - 2) = (x - 2)(x + 2)/(x - 2) = x + 2 = 2 + 2 = 4$, $x \neq 2$.

$\lim_{x \to 3} (-2x)/(x - 3)$ does not exist because right limit is $-\infty$, while left limit is $+\infty$.

Example:

Show that $f(x) = (1 - x^2)^{1/2}$ is continuous on closed interval $[-1, 1]$.

Function is continuous on closed interval $[a, b]$, where $a < b$, if it is continuous on open interval (a, b) and both the right limit at a and the left limit at b exist (continuous on endpoints a and b).

Function $f(x) = (1 - x^2)^{1/2}$ is continuous for all x as a polynomial function.

Limit as $x \to -1$ from the right = 0 [also, $f(-1)$ is defined, $f(-1) = 0$]
Limit as $x \to 1$ from the left = 0 [also, $f(1)$ is defined, $f(1) = 0$]

Thus, limits at endpoints -1 and 1 exist; $f(x)$ is continuous at endpoints a and b, and so it is continuous on closed interval $[-1, 1]$.

Example:

Find x-values for which $f(x) = x \div (x^2 - x)$ is not continuous. If there are any discontinuities, explain if they are removable or nonremovable.

F(x) is not continuous for x = 0 and 1 because at these values the denominator is zero.

Discontinuity x = 0 is removable because $\lim_{x \to 0} f(x) = -1$.
Discontinuity x = 1 is nonremovable because $\lim_{x \to 1} f(x) = +\infty$ (limit does not exist).

Example:

Determine if function $f(x) = (x^2 - 6x - 7)/(x + 1)$ has any discontinuities. If any discontinuity exists, explain whether it is removable or nonremovable.

The discontinuity at x = –1 is removable because $(x - 7)(x + 1)/(x + 1) = (x - 7)$. The $\lim_{\to -1} (x - 7) = -8$. Therefore, the line x = –1 is not an asymptote (removable).

Exercises:

Find the limits and discontinuities and classify them as removable/nonremovable:

1. $\lim_{x \to 0} (x^2 - x)/x$ [x – 1]

2. $\lim_{x \to 4} (x - 4)/(x^2 - 16)$ [1/8]

3. $f(x) = 3/(x - 2)$ [x = 2, nonremovable]

4. $f(x) = (x^2 - 9)/(x^2)$

5. $f(x) = x/(x^2 - x)$ [x = 0 is removable, x = 1 is nonremovable]

The concept of a limit is the fundamental building block of Calculus. Traditionally, Calculus is divided into Differential and Integral parts. The Differential part deals with concepts that arise from the tangent line problem and the Integral part deals with concepts that arise from the area problem.

CHAPTER 3 – DERIVATIVES

DERIVATIVE AS A LIMIT:

Consider two points on the graph of the function f(x) and the secant line QP

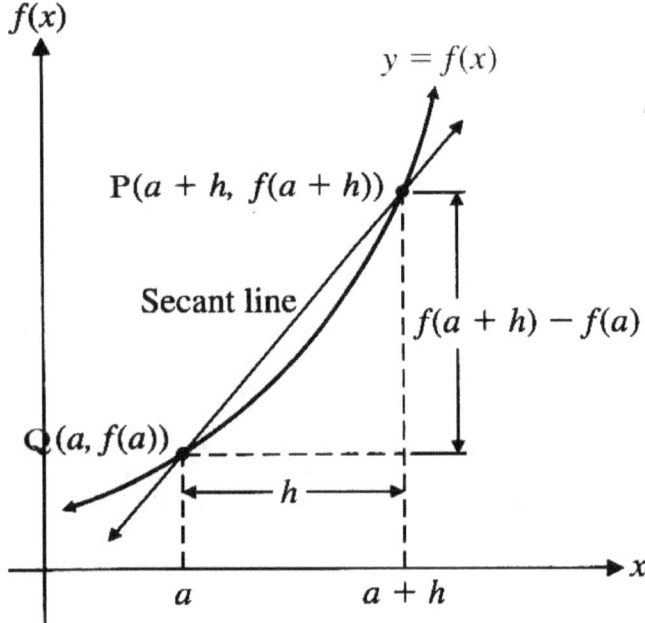

between them. The slope of the secant line is rise/run, or "change in y" divided by "change in x", which is [f(a + h) – f(a)]/h

This is the definition of the slope of the line between two points. Imagine that the point Q moves closer and closer to the point P. Then the secant line would get shorter and shorter until the points Q and P coincide, in which case the secant line between Q and P would be the tangent line to the graph of f(x) at the point P.

The definition of the derivative of the function f(x) at a given point P is the limit of the slope function of the secant line as the distance h between two points approaches zero.

$$\text{The Derivative} = f(x)' = \lim_{h \to 0} [f(a+h) - f(a)]/h$$

The expression [f(a + h) – f(a)]/h is called the "difference quotient". The derivative of f(x) at x is the limit of the difference quotient as h approaches zero. It is important to understand that the derivative of function is *defined* as a limit. It is the limit of slopes of secant lines drawn through points Q and P. As point Q moves closer and closer to its limiting position at point P, the distance (Δx) between x coordinates of Q and P becomes progressively smaller, as Δx approaches zero.

The distance between y coordinates of Q and P also approaches zero as the points coincide. Each y coordinate of a point (x, y) is simply the value of the function f(x), that is y = f(x). Therefore, every point (x, y) can be written as (x, f(x)). The slope of the line between two points (x_0, y_0) and (x_1, y_1) is m = $(y_1 - y_0)/(x_1 - x_0)$, therefore m = $(f(x_1) - f(x_0))/(x_1 - x_0)$. The coordinate x_1 is simply x_0 plus the distance ("change in x" or "Δx" or "dx") between x coordinates: $x_1 = x_0 + \Delta x$. Thus, the y coordinate is simply $f(x_1)$ which is $f(x_0 + \Delta x)$.

The formal definition of the derivative of f(x) is the limit of the distance between y values divided by the distance between x values (Δx), as Δx approaches zero ("change in x" is abbreviated by Greek letter "delta" - "Δx" or Latin letter "h")

$d/dx[f(x)] = \lim_{\Delta x \to 0} [f(x + \Delta x) - f(x)]/\Delta x$ or $f(x)' = \lim_{h \to 0} [f(x + h) - f(x)]/h$

The derivative of function f(x) may be interpreted as function whose value at point (x, y) is the slope of the tangent line to the graph of y = f(x) at point (x, y).

The derivative of function f(x) is another function, denoted $d/dx[f(x)]$ or $f'(x)$. The derivative of function at a point x may be interpreted as a slope to the graph of this function at x or rate of change at an instant of time. But the actual *definition* of the derivative is the limit of the difference quotient.

The notation for the derivative is $d/dx[f(x)]$ which means "derivative of f(x)" or dy/dx which means "derivative of y with respect to x". This is so called "Leibniz Notation" named after Gottfried Wilhelm Leibniz. Another notation for the derivative of function is called "Prime Notation" due to Isaac Newton: the derivative of f(x) is written $f(x)'$.

Example:

Find the derivative of $f(x) = 2x^2 + 1$ using limit definition of derivative.
The definition of derivative of f(x) is $d/dx[f(x)] = \lim_{h \to 0} [f(x + h) - f(x)]/h$

$f(x + h) = 2(x + h)^2 + 1 = 2(x^2 + 2xh + h^2) + 1 = 2x^2 + 4xh + 2h^2 + 1$
$f(x + h) - f(x) = 2x^2 + 4xh + 2h^2 + 1 - 2x^2 - 1 = 4xh + 2h^2$

Thus, $\lim_{h \to 0} [f(x + h) - f(x)]/h = \lim_{h \to 0} (4xh + 2h^2)/h = \lim_{h \to 0} (4x + 2h) = 4x$.

Example:

Find the derivative of $f(x) = 4x - x^2$ using limit definition of derivative.

$f(x + h) = 4(x + h) - (x + h)^2$ and $f(x) = 4x - x^2$. Therefore, the derivative is equal to $\lim_{h \to 0} [f(x + h) - f(x)]/h = \lim_{h \to 0} [4(x + h) - (x + h)^2 - (4x - x^2)]/h =$
$= \lim_{h \to 0} [4x + 4h - (x^2 + 2xh + h^2) - 4x + x^2]/h =$
$= \lim_{h \to 0} [4x + 4h - x^2 - 2xh - h^2 - 4x + x^2]/h =$
$= \lim_{h \to 0} [4h - 2xh - h^2]/h = \lim_{h \to 0} h[4 - 2x - h]/h = 4 - 2x.$

This new function $(4 - 2x)$ is the derivative of the original function $f(x) = 4x - x^2$. The derivative is usually denoted by $f'(x)$, read "f prime". The derivative function $f'(x)$ evaluated at a given value of x is the slope of the tangent line to the graph of $f(x)$ at the point $(x, f(x))$. For example, the slope of the tangent line to the graph of $f(x) = 4x - x^2$ when $x = 3$ is equal to $f'(3) = 4 - 2(3) = -2$. When $x = 3$, the value of the original function is equal to $f(3) = 4(3) - (3)^2 = 12 - 9 = 3$.

Example:

Find the derivative of $f(x) = x - 3x^2$ using the limit definition of derivative. Write an equation of the line tangent to $f(x)$ at $(2, -10)$.

Derivative of $f(x) = x - 3x^2$ is a limit of $[f(x + \Delta x) - f(x)]/\Delta x$ as Δx approaches zero:
$\lim_{\Delta x \to 0} [(x + \Delta x) - 3(x + \Delta x)^2 - (x - 3x^2)]/\Delta x =$
$= \lim_{\Delta x \to 0} [x + \Delta x - 3x^2 - 6x\Delta x - 3(\Delta x)^2 - x + 3x^2]/\Delta x =$
$= \lim_{\Delta x \to 0} [\Delta x - 6x\Delta x - 3(\Delta x)^2]/\Delta x = \lim_{\Delta x \to 0} [\Delta x(1 - 6x - 3\Delta x)]/\Delta x = 1 - 6x.$

The derivative is the slope at the point where $x = 2$, that is, $f'(2) = 1 - 12 = -11$. Then, $y + 10 = -11(x - 2)$ and therefore $y = -11x + 12$, which is an equation of tangent line at $(2, -10)$.

Exercises:

Find the derivative of a given function using the limit definition of derivative:

1. $f(x) = 5x$ [5]

2. $f(x) = x^2 + 1$ [2x]

3. $f(x) = 3x^2 - 2x - 5$ [6x - 2]

DERIVATIVES AND RULES OF DIFFERENTIATION:

By being able to compute derivatives, we could find the slope and the equation of a given line when we only have one point on this line – the point at which the line is tangent to the graph of function f(x).

The fundamental problem of Differential Calculus is to find the equation of a line – the special line that is tangent to the graph of function f(x) at the point P(x, y).

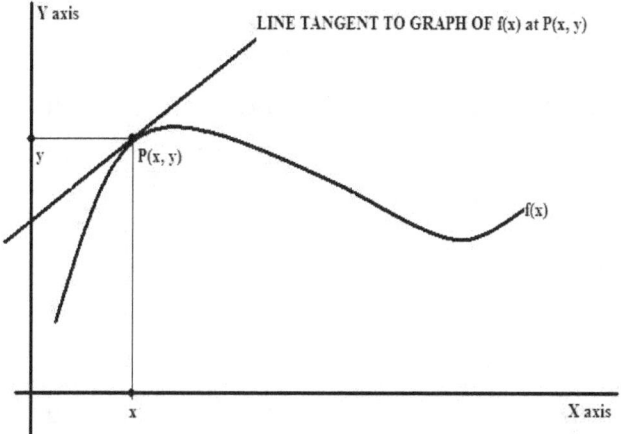

The Derivative of function f(x) is simply a formula for the slope of the line that is tangent to the graph of f(x) at the point P(x, y).

Rules of Differentiation:

Rule 1 The derivative of any real number (constant c) is zero: [c]′ = 0.

Examples:

d/dx[5] = 0, d/dx[− 7] = 0, d/dx[3/4] = 0

Derivative of a constant c times function f(x) is equal to constant times the derivative of this function: [c · f(x)]′= c · [f(x)]′. The derivative of x is equal to 1.

Examples:

d/dx[5x] = 5(x)′ = 5(1) = 5 (derivative of x is 1)
d/dx[− 7x] = − 7(x)′ = − 7(1) = − 7
d/dx[− 9(x + 2)] = − 9x − 18 = − 9(x)′ − 0 = − 9(1) = − 9.

Rule 2 Power Rule: the derivative of x with exponent n, where n is a real number, is given by the formula $[x^n]' = n \cdot x^{n-1}$.

Examples:

$d/dx[x^4] = 4x^3$
$d/dx[3x^5] = 5(3)(x^4) = 15x^4$
$d/dx[-2x^7] = 7(-2)(x^6) = -14x^6$
$d/dx[5x^{-10}] = -10(5)x^{-10-1} = -50x^{-11}$
$d/dx[5(3x^2 + x^7)] = 5(3(2)x^{2-1} + 7x^{7-1})] = 5(6x^1 + 7x^6) = 30x + 35x^6 = 5x(6 + 7x^6)$.

It is important to understand that to compute the derivative of function f(x) with exponent n we must multiply the exponent n *times* f(x) with exponent n − 1 and then *times* the derivative of f(x): $[f(x)n\ ]' = n \cdot [f(x)]^{n-1} \cdot [f(x)]'$

Example:

Find the derivative of $f(x) = (3x + 1)^5$.
The derivative of $(3x + 1)^5$ is exponent 5 *times* the function $(3x + 1)^{5-1}$ *times* the derivative of the function $(3x + 1)$:

$d/dx\ [(3x + 1)^5] = 5(3x + 1)^{5-1} \cdot d/dx[3x + 1] = 5(3x + 1)^4\ (3) = 15(3x + 1)^4$.

Note: the derivative of $(3x + 1)$ is 3.

Example:

Find the derivative of $f(x) = (x^2 − 5)^4$.
The derivative is $d/dx[(x^2 − 5)^4] = 4(x^2 − 5)^3(2x) = 8x(x^2 − 5)^3$.

Note: the derivative of $(x^2 − 5)$ is 2x.

Example:

Find the derivative of $f(x) = (x^3 + 5x^2 + 6x + 15)^7$.

$d/dx[(x^3 + 5x^2 + 6x + 15)^7] = 7(x^3 + 5x^2 + 6x + 15)^6(3x^2 + 10x + 6)$.

Note: the derivative of $(x^3 + 5x^2 + 6x + 15)$ is $(3x^2 + 10x + 6)$.

Example:

Find the derivative of $f(x) = (x^2 - 6x + 2)^5$.

$d/dx[(x^2 - 6x + 2)^5] = 5(x^2 - 6x + 2)^4(2x - 6)$.

Note: the derivative of $(x^2 - 6x + 2)$ is $(2x - 6)$.

Rule 3 The derivative of a sum of two functions is given by the formula
$[f(x) + g(x)]' = [f(x)]' + [g(x)]'$.

Example:

Let $f(x) = 3x + 1$ and $g(x) = x^5$, then $d/dx[f(x)] = 3$ and $d/dx[g(x)] = 5x^4$ and so
$d/dx[f(x) + g(x)] = d/dx[3x + 1] + d/dx[x^5] = 3 + 5x^4$

Rule 4 The derivative of a difference of two functions is given by the formula
$[f(x) - g(x)]' = [f(x)]' - [g(x)]'$.

Example:

Let $f(x) = 4x + 2$ and $g(x) = x^{10}$, then $d/dx[f(x)] = 4$ and $d/dx[g(x)] = 10x^9$.
Then $d/dx[f(x) - g(x)] = d/dx[4x + 2] - d/dx[x^{10}] = 4 - 10x^9$.

Rule 5 Product Rule: the derivative of a product of two functions is given by the formula
$(f \cdot g)' = (f' \cdot g) + (f \cdot g')$

That is, derivative of a product of two functions is the first function times the derivative of the second function *plus* the second function times the derivative of the first function. Three calculation steps:

1. Multiply the first function times the derivative of the second function
2. Multiply the second function times the derivative of the first function
3. Add the results of #1 and #2

Example:

Find the derivative of $f(x) \cdot g(x)$ if $f(x) = x + 4$ and $g(x) = 2x - 3$.

First, we need to find the derivatives of each function: d/dx[f(x)] = 1 and
d/dx[g(x)] = 2

1. The first function times the derivative of the second function is
(x + 4)(2) = 2x + 8
2. The second function times the derivative of the first function is
(2x − 3)(1) = 2x − 3
3. Adding #1 and #2: (2x + 8) + (2x − 3) = 4x + 5

The derivative of the product f(x) · g(x) is 4x + 5.

Example:

Find the derivative of the product f(x) · g(x) if f(x) = 3x + 1 and g(x) = x^5.

First, we need to find the derivatives of each function:
d/dx[f(x)] = 3 and d/dx[g(x)] = $5x^4$

1. First function times the derivative of the second function is
(3x + 1)($5x^4$) = $15x^5$ + $5x^4$
2. The second function times the derivative of the first function is (x^5)(3) = $3x^5$
3. Adding #1 and #2: ($15x^5$ + $5x^4$) + $3x^5$ = $18x^5$ + $5x^4$

The derivative of the product (3x + 1)(x^5) is $18x^5$ + $5x^4$.

Example:

Find the derivative of the product f(x) · g(x) if f(x) = 2x + 7 and g(x) = x − 9.
First, we need to find the derivatives of each function: d/dx[f(x)] = 2 and
d/dx[g(x)] = 1

1. The first function times the derivative of the second function is (2x + 7)(1)
2. The second function times the derivative of the first function is
(x − 9)(2) = 2x − 18
3. Adding #1 and #2: (2x + 7) + (2x − 18) = 4x − 11

The derivative of the product (2x + 7)(x − 9) is 4x − 11.

Example:

Find the derivative of the product $(3x^2 - 1)(7 + 2x^3)$.
The derivative of $(3x^2 - 1)$ is $6x$ and derivative of $(7 + 2x^3)$ is $6x^2$

1. First function times the derivative of the second function is
$(3x^2 - 1)(6x^2) = 18x^4 - 6x^2$

2. Second function times the derivative of the first function is
$(7 + 2x^3)(6x) = 42x + 12x^4$

3. Adding #1 and #2: $18x^4 - 6x^2 + 42x + 12x^4 = 30x^4 - 6x^2 + 42x$.

The derivative of the product $(3x^2 - 1)(7 + 2x^3)$ is $30x^4 - 6x^2 + 42x$.

Example:

Find the derivative of the product $(4x^2 - 1)(7x^3 + x)$.

The derivative of $(4x^2 - 1)$ is $8x$ and derivative of $(7x^3 + x)$ is $21x^2 + 1$

1. First function times the derivative of the second function is
$(4x^2 - 1)(21x^2 + 1) = 84x^4 + 4x^2 - 21x^2 - 1 = 84x^4 - 17x^2 - 1$

2. Second function times the derivative of the first function is
$(7x^3 + x)(8x) = 56x^4 + 8x^2$

3. Adding #1 and #2: $(84x^4 - 17x^2 - 1) + (56x^4 + 8x^2) = 140x^4 - 9x^2 - 1$

The derivative of the product $(4x^2 - 1)(7x^3 + x)$ is $140x^4 - 9x^2 - 1$.

Example:

Find the derivative of $f(x) = \sin x \cdot \cos x$.

The derivative of $\sin x$ is $\cos x$ and the derivative of $\cos x$ is $(-\sin x)$.
Therefore, $(\sin x \cdot \cos x)' = \sin x(-\sin x) + \cos x(\cos x) = -\sin^2 x + \cos^2 x$.

Rule 6 Quotient Rule: the derivative of a quotient $f(x) \div g(x)$ is given by the formula $(f \div g)' = [(f' \cdot g) - (f \cdot g')] \div (g)^2$.

That is, the derivative of the quotient of two functions is the denominator times the derivative of the numerator minus the numerator times the derivative of the denominator, divided by the denominator squared:

1. Multiply the denominator times the derivative of the numerator
2. Multiply the numerator times the derivative of the denominator
3. Subtract the results of #2 from #1: (#1 − #2)
4. Divide the result of #3 by the square of the denominator function

Example:

Find the derivative of $f(x) \div g(x)$ if $f(x) = 2x$ and $g(x) = x + 3$.

$d/dx[2x] = 2$ and $d/dx[x + 3] = 1$, $[g(x)]^2 = (x + 3)^2$

1. Multiply the denominator times the derivative of the numerator: $(x + 3)(2)$

2. Multiply the numerator times the derivative of the denominator: $(2x)(1)$

3. Subtract the results of #2 from #1: $(x + 3)(2) - (2x)(1) = 6$

4. Divide the result of #3 by the denominator squared: $6 \div (x + 3)^2$

The derivative of $[2x/(x + 3)]$ is $6/(x + 3)^2$.

Example:

Find the derivative of $f(x) \div g(x)$ if $f(x) = 4x - 1$ and $g(x) = x^2 + 3$.

$d/dx[4x - 1] = 4$ and $d/dx[x^2 + 3] = 2x$, $[g(x)]^2 = (x^2 + 3)^2$

1. Multiply the denominator times the derivative of the numerator:
$(x^2 + 3)(4) = 4x^2 + 12$

2. Multiply the numerator times the derivative of the denominator:
$(4x - 1)(2x) = 8x^2 - 2x$

3. Subtract the results of #2 from #1:
$(4x^2 + 12) - (8x^2 - 2x) = 4x^2 + 12 - 8x^2 + 2x = -4x^2 + 2x + 12$

4. Divide the result of #3 by the denominator squared: $(-4x^2 + 2x + 1) \div (x^2 + 3)^2$.

Example:

Find the derivative of $f(x) \div g(x)$ if $f(x) = x^3 + 4x^2$ and $g(x) = 5x$.

$d/dx[f(x)] = 3x^2 + 8x$ and $d/dx[g(x)] = 5$, $[g(x)]^2 = 25x^2$

1. The denominator times the derivative of the numerator is
$(5x)(3x^2 + 8x) = 15x^3 + 40x^2$

2. The numerator times the derivative of the denominator is
$(x^3 + 4x^2)(5) = 5x^3 + 20x^2$

3. Subtract the results of #2 from #1: $(15x^3 + 40x^2) - (5x^3 + 20x^2) = 10x^3 + 20x^2$

4. Divide the result of #3 by the square of the denominator:
$(10x^3 + 20x^2) \div 25x^2 = (10x^3) \div (25x^2) + (20x^2) \div (25x^2) = (2/5)x + 4/5$.

Example:

Find the derivative of $(x - 3) \div (x^2 + 7)$.

The derivative of $(x - 3)$ is 1 and the derivative of $(x^2 + 7)$ is $2x$

1. Multiply the denominator times the derivative of the numerator: $(x^2 + 7)(1)$

2. Multiply the numerator times the derivative of the denominator: $(x - 3)(2x)$

3. Subtract the results of #2 from #1: $(x^2 + 7) - (2x^2 - 6x) = -x^2 + 6x + 7$

4. Divide the result of #3 by the denominator squared: $(-x^2 + 6x + 7) \div (x^2 + 7)^2$.

Example:

Find the derivative of $(4x - 7) \div (3 - x^2)$.

The derivative of $(4x - 7)$ is 4 and the derivative of $(3 - x^2)$ is $-2x$

1. Multiply the denominator times the derivative of the numerator:
$(3 - x^2)(4) = 12 - 4x^2$

2. Multiply the numerator times the derivative of the denominator:
$(4x - 7)(-2x) = -8x^2 + 14$

3. Subtract the results of #2 from #1:
$(12 - 4x^2) - (-8x^2 + 14x) = 12 - 4x^2 + 8x^2 - 14x = 4x^2 - 14x + 12$

4. Divide the result of #3 by the denominator squared: $(4x^2 - 14x + 12) \div (3 - x^2)^2$.

Example:

Find the derivative of $(x^2 + x - 2) \div (x^3 + 4)$.

The derivative of $(x^2 + x - 2)$ is $2x + 1$ and the derivative of $(x^3 + 4)$ is $3x^2$

1. Multiply the denominator times the derivative of the numerator: $(x^3 + 4)(2x + 1)$

2. Multiply the numerator times the derivative of the denominator: $(x^2 + x - 2)(3x^2)$

3. Subtract the results of #2 from #1:
$(x^3 + 4)(2x + 1) - (x^2 + x - 2)(3x^2) = (2x^4 + x^3 + 8x + 4) - (3x^4 + 3x^3 - 6x^2) =$
$= -x^4 - 2x^3 + 6x^2 + 8x + 4$

4. Divide the result of #3 by the denominator squared:
$(-x^4 - 2x^3 + 6x^2 + 8x + 4) \div (x^3 + 4)^2$.

Example:

Find the derivative of $y = (\sec x)/x$.

The derivative of $\sec x$ is $(\sec x)(\tan x)$ and derivative of x is 1.
Therefore, $y' = [x(\sec x)(\tan x) - \sec x(1)]/x^2$.

Exercises:

Find the derivatives of the following functions:

1. $f(x) = x^2 + 5$ [2x]

2. $f(x) = 3x^2 + 5x + 1$

3. $f(x) = (x - 1)(x + 1)$ [2x]

4. $f(x) = (3x^2 - 5x - 1) + (x^3 - x^2 + 8x)$

5. $f(x) = x^4 + x^2$ [$4x^3 + 2x$]

6. $f(x) = (2x - 1)(3x^4 + 5x)$

7. $f(x) = (4x^2 - 1)(7x^3 + x)$ [$140x^4 - 9x^2 - 1$]

8. Find $d/dx(f(x)/g(x))$, $f(x) = x + 2$, $g(x) = x - 2$, $x \neq 2$

9. Find $d/dx(f(x)/g(x))$, $f(x) = (x^2 + 3)$, $g(x) = (5x - 1)^2$, $x \neq 1/5$

10. Find $d/dx(f(x))$ if $f(x) = (x^2 - 1)/(x^4 + 1)$ [$(-2x^5 + 4x^3 + 2x)/(x^4 + 1)^2$]

WRITING EQUATIONS OF LINES TANGENT TO FUNCTION F(X):

One of the most important problems of Differential Calculus is writing an equation of the line tangent to the graph of a given function f(x) at a given point (x_0, y_0).

But first, to develop the concept of the Derivative further, we need to review the basics of writing linear equations:

Example:

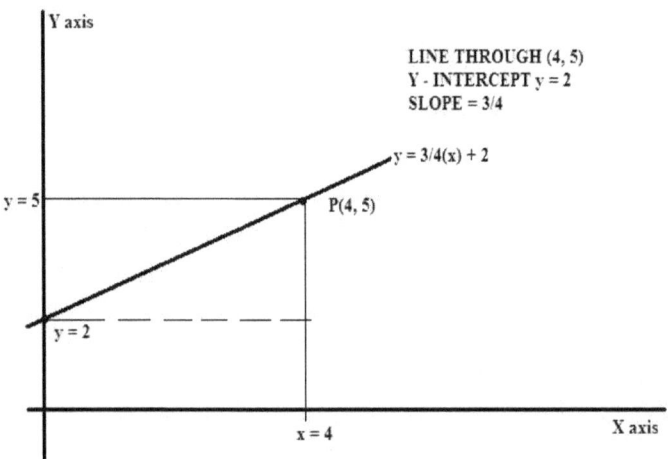

What is an equation of this line?

One formula for the linear equation is slope/y-intercept formula $y = mx + b$, where m is slope and b is y-intercept.

In this example slope is 3/4, which is simply the distance "up" (rise = 3) divided by the distance "across" (run = 4). The y-intercept is 2, which is the point where the line intersects y-axis. Therefore, $y = mx + b$ becomes $y = 3/4(x) + 2$. This is the equation of the line with slope $m = 3/4$ and y-intercept $y = 2$.

Another way to find an equation of the line when we do not have y-intercept is the point/slope formula $y - y_0 = m(x - x_0)$ where m is the slope of the line that passes through the point (x_0, y_0). In this example we have two points on the line: (0, 2) and (4, 5). That is, $y_2 = 5$, $y_1 = 2$, $x_2 = 4$, $x_1 = 0$. Therefore, the slope is $(5 - 2)/(4 - 0)$ or 3/4. This line passes through points (0, 2) and (4, 5), so we can select either point and use point/slope formula to obtain equation of this line. If we select point (4, 5), we obtain

$y - 5 = (3/4)(x - 4)$
$y - 5 = (3/4)(x) - (3/4)(4)$
$y - 5 = (3/4)(x) - 3$
$y = (3/4)(x) - 3 + 5$
$y = (3/4)(x) + 2$ or $y - (3/4)x = 2$.

Of course, both slope/intercept and point/slope formulas produce the same equation.

Example:

What is an equation of this line?

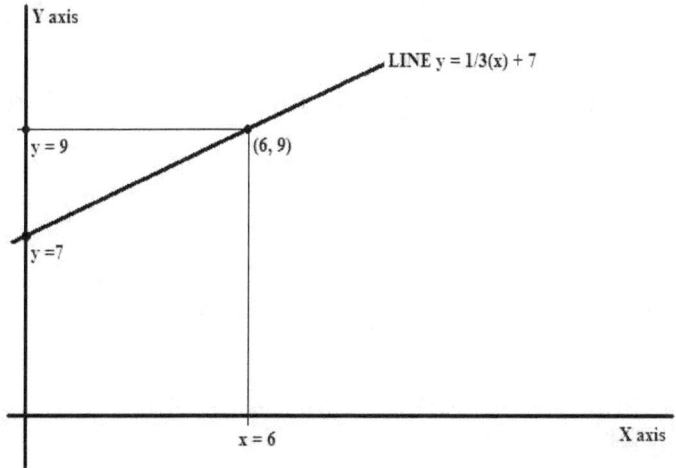

In this example, rise is $9 - 7 = 2$ and run is 6; so slope is $2/6 = 1/3$, while y-intercept is 7. Equation of the line is $y = (1/3)(x) + 7$.

We can also use the formula for the slope $m = (y_2 - y_1)/(x_2 - x_1)$. In this example we have two points on the line: $(0, 7)$ and $(6, 9)$. Thus, $y_2 = 9$, $y_1 = 7$, $x_2 = 6$, $x_1 = 0$. The slope is $(9 - 7)/(6 - 0) = 2/6 = 1/3$.

We can use either method to find slope:

1. Simply look at the picture of the graph to figure out "rise" and "run", and then use the definition: slope m = rise/run, or

2. Identify any two points (x_1, y_1) and (x_2, y_2) on the line and then use the formula for the slope $m = (y_2 - y_1)/(x_2 - x_1)$. If we do not know y–intercept but know slope

and at least one point on the line, we can use point/slope formula
$y - y_0 = m(x - x_0)$.

In the example, the line with the slope 1/3 is passing through the point (6, 9). Therefore,

$y - 9 = 1/3(x - 6)$
$y - 9 = 1/3(x) - 2$
$y = (1/3)x + 7$. Thus, the equation of the line is $y = (1/3)x + 7$.

Example:

Find an equation of a line through (2, 4) and (3, 21).

The slope is rise/run $= (21 - 4)/(3 - 2) = 17/1$. Once we have the slope, we can find an equation of the line using $y = mx + b$, where $x = 2$, $y = 4$ and $m = 17$. Therefore,

$4 = 17(2) + b$
$4 = 34 + b$
$4 - 34 = b$
$b = -30$ (that is, y–intercept is -30). Therefore, the equation is $y = 17x - 30$.

Equivalently, by using point/slope formula $y - y_0 = m(x - x_0)$, we obtain
$y - 4 = 17(x - 2) = 17x - 34$
$y = 17x - 34 + 4 = 17x - 30$.

Example:

Find the equation of a line through (2, 5) with slope 4.

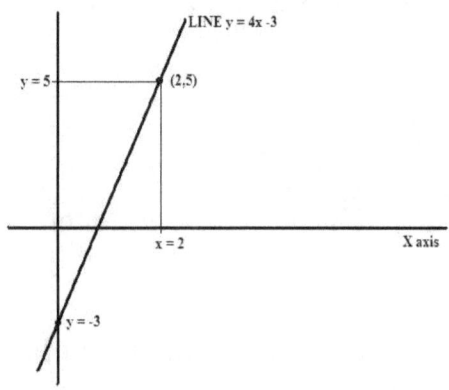

We can use the point/slope formula $y - y_0 = m(x - x_0)$ where $m = 4$, $x_0 = 2$, $y_0 = 5$:

$y - 5 = 4(x - 2) = 4x - 8$
$y = 4x - 8 + 5 = 4x - 3$

The equation of the line is $y = 4x - 3$.

Note: to write an equation of a line, we can use either slope/intercept or point/slope formula. We use slope/y-intercept formula $y = mx + b$ when we have point (x, y) on the line and both slope and y-intercept. If we have two points, or only have a slope and one point on the line then it is easier to use point/slope formula $y - y_0 = m(x - x_0)$.

To write an equation of the line tangent to the graph of a given function $f(x)$ at a given point (x_0, y_0), we need to complete three steps:

1. Find the derivative of the function $f(x)$.
2. Evaluate the derivative at the point (x, y) where the line is tangent to the graph of $f(x)$. This derivative is the slope of the tangent line at the point (x, y).
3. Write the equation of the tangent line by using point/slope or slope/intercept formula.

Example:

Write an equation of the line tangent to the graph of $f(x) = x^2$ at the point $(2, 4)$.

To write this equation we need to do three things:

1. Find the derivative of the function $f(x)$.
2. Evaluate the derivative at $x = 2$. This is the slope.
3. Use slope/y-intercept or point/slope formula to write an equation.

1. The derivative of the function $f(x) = x^2$ is $2x$ (Rule #2).

2. To find slope, we need to evaluate $2x$ at $x = 2$, which is 4. That is, slope $m = 4$.

3. Once we have the slope, we can find the equation of the line using slope/y-intercept formula $y = mx + b$, where $y = 4$, $x = 2$, $m = 4$. Therefore,

$4 = 4(2) + b$
$4 = 8 + b$

$4 - 8 = b$
$b = -4$.

That is, y-intercept is $y = -4$ and therefore the equation is $y = 4x - 4$. Equivalently, we can use point/slope formula $y - y_0 = m(x - x_0)$ with $m = 4$, $y = 4$, $x = 2$ equation is $y - 4 = 4(x - 2) = 4x - 8$ or $y = 4x - 8 + 4 = 4x - 4$, $y - 4x = 4$ or $y = 4x - 4$.

Therefore, an equation of the line tangent to $f(x) = x^2$ at $(2, 4)$ is $y = 4x - 4$.

Example:

Write an equation of the line tangent to the graph of the function $f(x) = x^2 + 1$ at the point $(2, 5)$. To write this equation we need to do three things:

1. Find the derivative of the function $f(x)$.
2. Evaluate the derivative at $x = 2$. This is the slope.
3. Use slope/y–intercept or point/slope formula to write an equation.

1. The derivative of the function $f(x) = x^2 + 1$ is $2x$ (Rules #2 and #1).

2. To find slope, we need to evaluate $2x$ at $x = 2$ which is 4.
Therefore, slope $m = 4$.

3. Once we have the slope, we can find the equation of the line using slope/y–intercept formula $y = mx + b$, where $y = 5$, $x = 2$, $m = 4$. Therefore,

$5 = 4(2) + b$
$5 = 8 + b$
$5 - 8 = b$
$b = -3$. That is, y-intercept is $y = -3$ and equation is $y = 4x - 3$.

Equivalently, we can use point/slope formula $y - y_0 = m(x - x_0)$ with $x_0 = 2$ and $y_0 = 5$.
$y - 5 = 4(x - 2) = 4x - 8$
$y = 4x - 8 + 5 = 4x - 3$.

Thus, an equation of the line tangent to $f(x) = x^2 + 1$ at $(2, 5)$ is $y = 4x - 3$.

Example:

Write an equation of the line tangent to the graph of the function $f(x) = x^3 + 5x$ at (1, 6).

To write this equation we need to do three things:

1. Find the derivative of the function f(x).
2. Evaluate the derivative at x = 1. This is the slope.
3. Use slope/y-intercept or point/slope formula to write an equation.

1. The derivative of the function $x^3 + 5x$ is $3x^2 + 5$ (Rules #1, #2 and #3).

2. To find slope, we need to evaluate $3x^2 + 5$ at x = 1, thus $3(1)^2 + 5 = 3(1) + 5 = 8$. Therefore, slope is 8.

3. Once we have the slope, we can find the equation of the line using slope/y-intercept formula y = mx + b, where y = 6, x = 1, m = 8. Therefore,

6 = 8(1) + b
6 = 8 + b
6 − 8 = − 2
Thus b = − 2 and equation is y = 8x − 2.

Equivalently, by using point/slope formula $y - y_0 = m(x - x_0)$, we obtain
y − 6 = 8(x − 1) = 8x − 8
y = 8x − 8 + 6 = 8x − 2.

Thus, an equation of the line tangent to $f(x) = x^3 + 5x$ at (1, 6) is y = 8x − 2.

Example:

Write an equation of the line tangent to $f(x) = 3x^2 - 6x + 4$ at the point (0, 4).

To write this equation we need to do three things:

1. Find the derivative of the function f(x).
2. Evaluate the derivative at x = 0. This is the slope.
3. Use slope/y-intercept or point/slope formula to write an equation.

1. The derivative of $f(x) = 3x^2 - 6x + 4$ is $6x - 6$ (Rules #1 and #3)

2. Evaluating $6x - 6$ at $x = 0$ we obtain $6(0) - 6 = -6$. Thus, slope $m = -6$.

3. The equation of the line is $y = mx + b$, which is $4 = -6(0) + b$, so that $b = 4$. Therefore, y-intercept is 4 and equation is $y = -6x + 4$.

Equivalently, using point/slope formula we obtain $y - 4 = -6(x - 0) = -6x$ or $y = -6x + 4$, which is an equation of the line tangent to $f(x)$ at the point $(0, 4)$.

Example:

Write an equation of the line tangent to $f(x) = -2x^3 + x^2 - 4x + 3$ at $(1, -2)$.

To write this equation we need to do three things:

1. Find the derivative of the function $f(x)$.
2. Evaluate the derivative at $x = 1$. This is the slope.
3. Use slope/y-intercept or point/slope formula to write an equation.

1. The derivative of $f(x) = -2x^3 + x^2 - 4x + 3$ is $-6x^2 + 2x - 4$ (Rules #1, #2 and #3)

2. Evaluating $(-6x^2 + 2x - 4)$ at $x = 1$, we obtain $-6(1)^2 + 2(1) - 4 =$
$= -6 + 2 - 4 = -8$.

3. An equation of the line is $y = mx + b$, $y = -2$, $x = 1$, $m = -8$. Therefore,
$-2 = -8(1) + b$
$-2 = -8 + b$
$-2 + 8 = b$

Thus, $b = 6$ and equation is $y = -8x + 6$.

Equivalently, $y + 2 = -8(x - 1) = -8x + 8$ or $y = -8x + 8 - 2 = -8x + 6$.
Thus, an equation of the line tangent to $f(x) = -2x^3 + x^2 - 4x + 3$ at point $(1, -2)$ is $y = -8x + 6$.

Example:

Write an equation of the line tangent to $h(x) = (2x)/(x + 3)$, $x \neq -3$ at $(3, 1)$.

To write this equation we need to do three things:

1. Find the derivative of the function $h(x)$.
2. Evaluate the derivative at $x = 3$. This is the slope.
3. Use slope/y-intercept or point/slope formula to write an equation.

Since $h(x)$ is the quotient of two other functions $f(x) = 2x$ and $g(x) = x + 3$, we must use the Quotient Rule to find the derivative of $h(x)$. We begin by finding the derivatives of $f(x)$ and $g(x)$, and $[g(x)]^2$: $d/dx[2x] = 2$, $d/dx[x + 3] = 1$, and $[g(x)]^2 = (x + 3)^2$.

1. Multiply the denominator times the derivative of the numerator: $(x + 3)(2)$

2. Multiply the numerator times the derivative of the denominator: $(2x)(1)$

3. Subtract the results of #2 from #1: $(x + 3)(2) - (2x)(1) = 6$

4. Divide the result of #3 by the denominator function squared: $6/(x + 3)^2$

The derivative of $h(x) = (2x)/(x + 3)$ is $6/(x + 3)^2$.

Evaluating $6/(x + 3)^2$ at the point where $x = 3$ we obtain the slope m of the line tangent to the graph of $h(x)$ at that point: Slope $m = 6/(3 + 3)^2 = 6/36 = 1/6$.

Using the point/slope formula $y - y_0 = m(x - x_0)$ where $m = 1/6$, $x_0 = 3$ and $y_0 = 1$ we obtain $y - 1 = 1/6(x - 3)$. Solving for y we obtain $y - 1 = x/6 - 3/6 = x/6 - 1/2$ $y = x/6 - 1/2 + 1$ and so $y = x/6 + 1/2 = (1/6)x + 1/2$. Therefore, an equation of the line tangent to $h(x) = (2x)/(x + 3)$ at the point $(3, 1)$ is $y = (1/6)x + 1/2$.

Example:

Find the slope of the line tangent to the function $f(x) = (3x^3 + 4x)^{1/5}$ at $x = 2$.

The slope of the tangent line at $x = 2$ is the derivative of $f(x)$ evaluated at $x = 2$, which is $f'(x) = (1/5)(3x^3 + 4x)^{-4/5}(9x^2 + 4) = (9x^2 + 4)/5(3x^3 + 4x)^{4/5}$. When evaluated at $x = 2$, the derivative $f'(2) = (9(2)^2 + 4)/5(3(2)^3 + 4(2))^{4/5} = 1/2$.

Example:

Write an equation of the line tangent to $f(x) = \tan x$ at the point $(\pi/4, 1)$.)
The derivative of $f(x) = \tan x$ is $f'(x) = \sec^2 x$. Evaluated at $x = \pi/4$, the derivative $f'(\pi/4) = \sec^2(\pi/4) = 2$, which is the slope of the tangent line at the point $(\pi/4, 1)$.

Equation of the tangent line is $y - 1 = 2(x - \pi/4) = 2x - \pi/2$. Thus, $y = 2x - \pi/2 + 1$.

Example:

Write an equation of the line tangent to $f(x) = 3\cos 2x - \sin x$, where $x = \pi/7$.

The derivative of $f(x) = 3\cos 2x - \sin x$ is $f'(x) = 3(-\sin 2x)(2) - \cos x$.
Evaluated at $x = \pi/7$, $f'(\pi/7) = -6\sin 2\pi/7 - \cos \pi/7 \approx -5.5919 \approx -5.6$, which is the slope of the tangent line at the point where $x = \pi/7$.

The y-value of this point is equal to $f(\pi/7) = 3\cos 2\pi/7 - \sin \pi/7 \approx 1.4365 \approx 1.4$.
Thus, point $(\pi/7, 1.4)$ is the point of tangency. The equation of the tangent line at this point is $y - 1.4 = -5.69(x - \pi/7)$, that is, $y = -5.69x + 5.69\pi/7 + 1.4$.

Example:

Find the point where $f(x) = x^{1/3}$ has a vertical tangent line.

Vertical line has an undefined slope. The derivative $f'(x) = (1/3)x^{-2/3} = 1/(3x)^{2/3}$.
The derivative is undefined for $x = 0$. Therefore, the function $f(x) = x^{1/3}$ has vertical tangent line at $x = 0$. The y-coordinate of this point is $y = f(0) = 0^{1/3} = 0$.

Therefore, $f(x) = x^{1/3}$ has a vertical tangent line at the point $(0, 0)$.

Example:

Find the point where $f(x) = x^2 + 9$ has a horizontal tangent line. Write an equation of this line at this point.

Horizontal line has a slope equal to zero. The derivative of $f(x) = x^2 + 9$ is $2x$.
The derivative is zero when $2x = 0$ and so $x = 0$. At $x = 0$, $f(0) = 9$.
Thus, the function $f(x) = x^2 + 9$ has a horizontal tangent line at the point $(0, 9)$.

The equation of this tangent line is $y - 9 = 0(x - 0)$, which is $y = 9$.

Example:

Find the point(s) where $f(x) = x^4 - 2x^2 + 3$ has a horizontal tangent line.

The derivative $f'(x) = 4x(x^2 - 1) = 4x(x - 1)(x + 1)$.
The derivative is equal to zero for $x = 0$, $x = -1$, and $x = 1$.
The corresponding values of $y = f(x)$ are $f(0) = 3$, $f(-1) = 2$, and $f(1) = 2$.

Thus, the points at which the function has a horizontal tangent line are $(0, 3)$, $(-1, 2)$, and $(1, 2)$.

Example:

Write an equation of the line tangent to $y = (x - 8)^{1/3}$ at the point $(8, 1)$.

Differentiating, we obtain $y' = (1/3)(x - 8)^{-2/3} = (1/3)/(x - 8)^{2/3}$.

At the point $x = 8$, $y' = (1/3)/(8 - 8)^{2/3}$ is undefined. Since $y'(8)$ is not defined, there is a vertical tangent line at the point $(8, 1)$.

Note: writing an equation of the line tangent to the graph of function $f(x)$ at a given point (x_0, y_0) is one of the central problems of Differential Calculus.

Exercises:

1. Find the equation of a line through $(0, 5)$ with slope 2. [$y = 2x + 5$]

2. Find the equation of a line through $(3, 6)$ with the y-intercept 4.

3. Find the equation of a line through $(6, 3)$ with the y-intercept 1.

4. Find the equation of a line through $(2, 4)$ and $(-1, -5)$. [$y = 3x - 2$]

5. Find the equation of a line through $(0, 0)$ with the slope 1.

6. Find the equation of a line through $(0, 3)$ and $(2, 9)$.

7. Find the equation of a line through $(4, 0)$ and $(4, 3)$. [$x = 4$]

AVERAGE AND INSTANTANEOUS VELOCITY

Velocity of an object is the rate of change of position with time. That is, how fast an object is moving means how fast the distance is changing in a given unit of time.

Average velocity describes how fast an object is moving during a given interval of time $[t_1, t_2]$. Average velocity is equal to distance traveled divided by the time it took to travel this distance. That is, $V(average) = D/t = [D(t_2) - D(t_1)]/(t_2 - t_1)$.

For example, suppose that the distance covered by a moving object is given by the formula $D(t) = 3t + 5$, where t is time in seconds. The average velocity of this object between 6th and 8th second is $[D(8) - D(6)]$ divided by $(8 - 6)$, therefore the average velocity is

$$V(average) = [(3(8) + 5) - (3(6) + 5)] \div (8 - 6) = [29 - 23] \div 2 = 3.$$

Instantaneous velocity is the speed of an object during an instant. The question here is not how fast an object moved during an interval of time, but how fast an object is moving at a given instant. Instantaneous velocity is the derivative of the distance function D(t) evaluated at an instant of time t.

For example, suppose that the distance covered by a moving object is given by the formula $D(t) = t^2 + 7t$, where distance is in feet and t is time in seconds. Then the instantaneous velocity at 3rd second is the derivative of distance, $D'(t) = 2t + 7$. This derivative evaluated at the 3rd second is $D'(3) = 2(3) + 7 = 13$ feet per second.

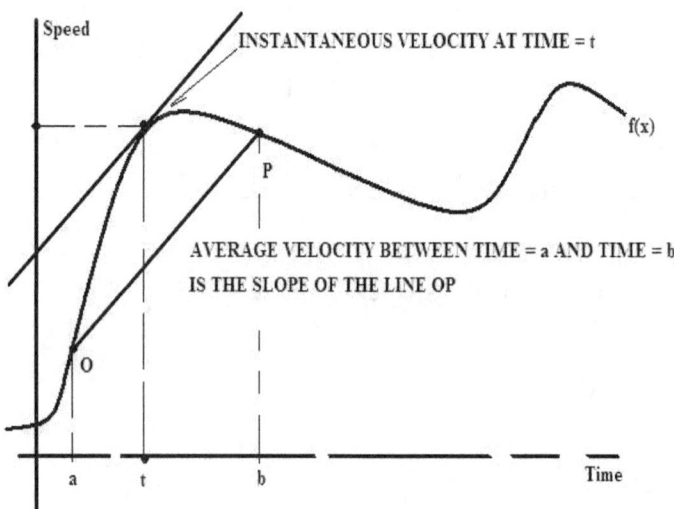

Example:

An object is moving along a line. The distance it has traveled is given by the formula $D(t) = t^2 - 6t + 8$, where distance is in feet and time is in seconds. What is the instantaneous velocity at $t = 4$?

The instantaneous velocity is the derivative of the distance function, which is $D'(t) = 2t - 6$. Evaluating $D'(t)$ at $t = 4$ we obtain $D'(4) = 2(4) - 6 = 2$. The instantaneous velocity at the 4th second is 2 feet per second.

Example:

An object is moving along a line. The distance it has traveled is given by the function $D(t) = 2t^2 - 10t + 13$, where distance is in feet and time is in seconds. Find the average velocity between 4th and 5th second and the instantaneous velocity precisely at 5th second.

Average velocity is $D/t = [D(5) - D(4)]/(5 - 4)$

$D(5) = 2(5^2) - 10(5) + 13 = 13$
$D(4) = 2(4^2) - 10(4) + 13 = 5$

Therefore, $D/t = [D(t_2) - D(t_1)]/(t_2 - t_1) = [13 - 5]/1 = 8$ feet per second (f/s), which is the average velocity between 4th and 5th second. The instantaneous velocity is the derivative of $D(t) = 2t^2 - 10t + 13$.

The derivative of the distance function is $D'(t) = 4t - 10$.
At the 5th second, the derivative is $D'(5) = 4(5) - 10 = 10$ f/s.

Example:

An object is moving along a line. The distance it has traveled is given by the formula $D(t) = t^2 - 4t + 7$, where distance is in feet and time is in seconds. Find the average velocity between 3rd and 5th second and the instantaneous velocity precisely at 3rd second.

Average velocity is $D/t = [D(5) - D(3)]/(5 - 3)$

$D(5) = 5^2 - 4(5) + 7 = 12$ and $D(3) = 3^2 - 4(3) + 7 = 4$

Therefore, D/t = [D(t$_2$) − D(t$_1$)]/(t$_2$ − t$_1$) = [12 − 4]/2 = 8 feet per second, which is the average velocity between 3 and 5 seconds.

The instantaneous velocity is the derivative of D(t) = t^2 − 4t + 7, evaluated at t = 3. The derivative of distance function is D'(t) = 2t − 4, which is D'(3) = 2(3) − 4 = 2. Therefore, the instantaneous velocity at the 3rd second is 2 feet per second.

Instantaneous acceleration is the rate of change of velocity with respect to time during an instant. Instantaneous acceleration is a measure of how fast the velocity is changing at a given instance. Instantaneous acceleration is the derivative of instantaneous velocity or the second derivative of the distance function:
A(t) = V'(t) = d/dt[V(t)] or D''(t) = d/dt[d/dt[D(t)]].

Example:

Let D(t) = t^3 − 6t^2 , find the instantaneous acceleration at t = 5.

Instantaneous velocity is V(t) = d/dt[D(t)] = d/dt[t^3 − 6t^2] = 3t^2 − 12t
Instantaneous acceleration is d/dt[V(t)] = d/dt[3t^2 − 12t] = 6t − 12

At t = 5, 6t − 12 = 6(5) − 12 = 18 feet per second squared (f/s^2). The instantaneous acceleration at the 5th second is 18 feet per second squared.

Example:

Let D(t) = 2t^3 − 21t^2 + 60t + 3, find the instantaneous acceleration at t = 4.

Instantaneous velocity V(t) is the derivative of the distance function d/dt[D(t)], which is d/dt[2t^3 − 21t^2 + 60t + 3] = 6t^2 − 42t + 60.

Instantaneous acceleration is A(t) = d/dt[V(t)] is d/dt[6t^2 − 42t + 60] = 12t − 42. At t = 4, instantaneous acceleration is A(4) = 12(4) − 42 = 6 f/s^2.

Example:

The distance formula is D(x) = x^4 − 6x^2 + 2. Find the instantaneous velocity and acceleration at x = 2 (2nd second).

The first derivative D'(t) = 4x^3 − 12x is instantaneous velocity. At the 2nd second, instantaneous velocity is $4(2)^3 - 12(2) = 8$ f/s. The second derivative is $12x^2 - 12$. At the 2nd second, the acceleration is $12(2)^2 - 12 = 36$ f/s^2.

Example:

Position of a moving particle is given by the function $P(t) = (1/2)t^3 - 2t$, where distance is in feet and t is time in seconds. Find instantaneous velocity and instantaneous acceleration at the 2nd second.

Instantaneous velocity at $t = 2$ is $P'(2) = (3/2)(2)^2 - 2 = 4$ ft/sec.
Instantaneous acceleration at $t = 2$ is $P''(2) = 3(2) = 6$ ft/sec^2.

Exercises:

1. An object moves along a straight line. Its distance formula is $D(t) = (t + 2)^2$. What is the formula for instantaneous velocity? [$2(t + 2)$]

2. An object moves along a straight line. Its distance formula is $D(t) = (3t + 3)^3$. What is the formula for instantaneous velocity?

3. The distance traveled by an object is given by $D(t) = t^3 - t^2$. What is its instantaneous velocity and instantaneous acceleration at $t = 1$?

4. The distance traveled by an object is given by $D(t) = t^3 - 6t^2$. What is its instantaneous velocity and acceleration at $t = 1$? [-9 ft/sec, -6 ft/sec^2]

5. The distance traveled by an object is given by $D(t) = 3t^2 + 3t - 6$. What is its instantaneous velocity and instantaneous acceleration at $t = 5$?

6. The position of a moving body is given by the equation $D(t) = 160t - 16t^2$. Find instantaneous velocity and instantaneous acceleration functions.

7. The position of a moving body is given by the equation $D(t) = t^3 - 9t^2 + 24t$. Find instantaneous acceleration formula. [$6(t - 3)$]

8. An object is moving along a line. Its position at time t is given by the formula $D(t) = t/(t^2 + 4)$. Find its instantaneous velocity and acceleration at $t = 2$.

CHAIN RULE:

The composition of two functions f(x) and g(x) is another function, written f(g(x)). The Chain Rule is used to find the derivative of such composition function.

Let's say we have a composition function f(g(x)). This composition function is really two functions: f(g) and g(x). To find the derivative of composition function f(g(x)), we need to find the derivative of function f(g), which is derivative of f with respect to g, and then find the derivative of function g(x), which is derivative of g with respect to x, and then multiply both derivatives: df/dx = df/dg · dg/dx.

Equivalently, having two functions f(x) and g(x), we can write a composition function f(g(x)) and then differentiate it directly using the formula
f'(g(x)) = f'(g(x)) · g'(x).

Formally speaking, let y = f(u) be a differentiable function of u and let u = g(x) be a differentiable function of x. Then y is also differentiable function of x and therefore we can write a new composition function y = f(g(x)).

The derivative of this function is y' = f'(g(x)) · g'(x).

This is the same as Chain Rule formula *dy/dx = (dy/du)·(du/dx)*

Example:

Let function $y(w) = w^5$ and function $w(x) = x^3 + 1$. That is to say, we have a composition function y(w(x)).

The derivative is of y is $y' = 5w^4$ and the derivative of function w is $w' = 3x^2$.

Multiplying the derivatives we obtain $(y')(w') = 5w^4 (3x^2)$.

Since $w = x^3 + 1$, we substitute $(x^3 + 1)$ instead of w into function $5w^4 (3x^2)$, thus obtaining $5(x^3 + 1)^4(3x^2)$, which is equal to $15x^2(x^3 + 1)^4$.

Equivalently, instead of having two functions $y(w) = w^5$ and $w(x) = x^3 + 1$, we can immediately write composition function $y(w(x)) = (x^3 + 1)^5$.

Then, by differentiating $y = (x^3 + 1)^5$, we obtain
$y' = 5(x^3 + 1)^4(3x^2) = 15x^2(x^3 + 1)^4$.

Example:

Suppose that sales (S) of a product depend on the amount of money (A) spend on advertising (A) whereas the amount of money spend on advertising depends on the previous year's profit (p). Then the sales ultimately depend on the previous year's profit. Thus, we have a composition function S(A(p)). Find the derivative of this composition function with respect to p where sales formula is $S = 4A^2 + 6A + 3$ and advertising formula is $A = 5p - 2$, where p is profit.

Solution (1):

1. Find the derivative of S(A) and the derivative of A(p).
2. Multiply these derivatives according to Chain Rule formula
dy/dx = (dy/du)·(du/dx)
3. Substitute the original function A(p) into #2.

dS/dp[S(A(p))] = dS/dA · dA/dp, which is read "derivative of S with respect to p is equal to derivative of S with respect to A *times* derivative of A with respect to p".

Let $S = 4A^2 + 6A + 3$ and let $A = 5p - 2$

Derivative of S is dS/dA = 8A + 6 and derivative of A is dA/dp = 5.

Following Chain Rule, we multiply these derivatives: dS/dA · dA/dp = (8A + 6)(5), which is 40A + 30.

Now, we substitute the original function A = 5p - 2 into 40A + 30, obtaining derivative of S(A(p)), which is 40(5p - 2) + 30 = 200p - 80 + 30 = 200p - 50.

Thus, the derivative of the composition function S(A(p)), read "derivative of S with respect to p", or "dS/dp", is equal to dS/dA · dA/dp, which is 200p - 50.

Solution (2):

This procedure can be much shorter if we substitute function A(p) = 5p - 2 into function $S(A) = 4A^2 + 6A + 3$, thus obtaining the function $S(A(p)) = 4(5p - 2)^2 + 6(5p - 2) + 3$. Differentiating S(A(p)), we obtain
S'(A(p)) = 4(2)(5p - 2)(5) + 6(5) = 40(5p - 2) + 30 = 200p - 80 + 30 = 200p - 50.

Note: Differentiating the composition function directly is often a shorter way.

Example:

Let $f(x) = x^2$ and $x(q) = 3q + 4$. Then the composition function $f(x(q)) = (3q + 4)^2$, which is $9q^2 + 24q + 16$. Differentiating this composition function directly we obtain $f'(x(q)) = 18q + 24$.

Equivalently:

1. Find the derivative of $f(x)$ and the derivative of $x(q)$.
2. Multiply these derivatives.
3. Substitute the original function $x(q)$ into step #2.

The derivative of the function f with respect to q would be the product of the derivatives of the function f with respect to x *times* the derivative of x with respect to q: $d/dx[f(x)] = d/dx[x^2] = 2x$ and $d/dx[x(q)] = d/dx[3q + 4] = 3$.

Multiplying these derivatives we obtain $(2x)(3) = 6x$.
Substituting $x = 3q + 4$ into $6x$ we obtain $6(3q + 4) = 18q + 24$.

Example:

Find the derivative with respect to x of $f(g) = g^4$, where g is $g(x) = 12x + 9$.

We can easily solve this problem by simply differentiating the composition function $f(g(x)) = (12x + 9)^4$ directly: the derivative of $f(g(x))$ is $f'(g(x)) = 4(12x + 9)^3(12) = 48(12x + 9)^3$.

Equivalently:

The derivative of $f(g)$ with respect to g is $d/dg\,[f(g)] = 4g^3$
The derivative of $g(x)$ with respect to x is $d/dx\,[g(x)] = 12$

The derivative of $f(g(x))$ with respect to x is the product of the derivative of $f(g)$ with respect to g and the derivative of $g(x)$ with respect to x:
$d/dg\,[f(g)] \cdot d/dx\,[g(x)]$, which is $(4g^3)(12)$.

Substituting $g(x) = 12x + 9$ into $(4g^3)(12)$ we obtain function
$4(12x + 9)^3(12) = 48(12x + 9)^3$.

Example:

Find the derivative of the function $Y(u) = u^2 - 6$ with respect to x, where $u(x) = 5x^2 + 1$.

We can easily solve this problem by simply differentiating $Y(u(x))$ directly:
$Y'(u(x)) = (u^2 - 6)' = [(5x^2 + 1)^2 - 6]' = 2(5x^2 + 1)(10x) = 100x^3 + 20x$.

Equivalently, since Y is function of u and u is function of x, we need the derivative of a composition function $Y(u(x))$. In other words, we need to

1. Find the derivative of $Y(u)$ and the derivative of $u(x)$.
2. Multiply these derivatives.
3. Substitute the original function $u(x)$ into step #2.

The derivative of $Y(u)$ is $dY/du = 2u$ and the derivative of $u(x)$ is $du/dx = 10x$.
The product of the derivatives is $dY/dx = (2u)(10x)$
Since $u(x) = 5x^2 + 1$, we can replace $u(x)$ with $5x^2 + 1$ and obtain
$dY/dx = 2(5x^2 + 1)(10x) = (10x^2 + 2)(10x) = 100x^3 + 20x$.

Example:

Find the derivative of function $f(u) = u^3 - 3u^2 + 1$ with respect to x, where $u(x) = x^2 + 2$.

We can easily solve this problem by simply differentiating $f(u(x))$ directly:
$f'(u(x)) = [u^3 - 3u^2 + 1]' = [(x^2 + 2)^3 - 3(x^2 + 2)^2 + 1]' = 3(x^2 + 2)^2(2x) - 3(2)(x^2 + 2)(2x) = 6x(x^2 + 2)^2 - 12x(x^2 + 2) = 6x(x^2 + 2)(x^2 + 2 - 2) = 6x^3(x^2 + 2)$.

Equivalently:

1. Find the derivative of $f(u)$ and the derivative of $u(x)$.
2. Multiply these derivatives.
3. Substitute the original function $u(x)$ into #2.

The derivative of $f(u)$ is $3u^2 - 6u$ and derivative of $u(x)$ is $2x$. Multiplying the derivatives we obtain $(3u^2 - 6u)(2x)$. Substituting the function $u(x) = x^2 + 2$ into $(3u^2 - 6u)(2x)$, we obtain $(3(x^2 + 2)^2 - 6(x^2 + 2))(2x)$. Thus, we get
$(3(x^4 + 4x^2 + 4) - 6x^2 - 12)(2x) = (3x^4 + 6x^2)(2x) = 6x^5 + 12x^3$.
Factoring out $6x^3$ we get $6x^3(x^2 + 2)$, the derivative of $f(u)$ with respect to x.

Example:

Find y' if $y = 1/(4x^2 + 6x - 7)^3$.

To simplify, we can represent y with a negative exponent: $y = (4x^2 + 6x - 7)^{-3}$.

Thus, $y' = -3(4x^2 + 6x - 7)^{-4}(4x^2 + 6x - 7)' = -3(4x^2 + 6x - 7)^{-4}(8x + 6)$.

Therefore, $y' = -6(4x + 3)/(4x^2 + 6x - 7)^4$.

Example:

Let $y = 4\sin(3x)$. This is a composition function $f(g(x))$. The derivative of y with respect to x is $y' = 4\cos(3x)(3) = 12\cos(3x)$. *Note:* derivative of sine is cosine and derivative of 3x is 3.

Example:

Let $y = \sec^3(5x)$. This is a composition function $f(g(x))$. The derivative of y with respect to x is $y' = 3[\sec(5x)]^2[\sec(5x)\tan(5x)(5)] = 15\sec^3(5x)\tan(5x)$. *Note:* derivative of sec(x) is sec(x)tan(x) and derivative of 5x is 5.

Example:

Write an equation of the line tangent to the graph of $f(x) = (4x^3 + 3)^2$ at $(-1, 1)$.

$y' = 2(4x^3 + 3)(12x^2) = 24x^2(4x^3 + 3)$. Evaluated at $x = -1$, $y' = -24$ and so an equation of the line tangent to f(x) at $(-1, 1)$ is $y - 1 = -24(x + 1)$, which is $y = -24x - 23$.

Example:

Write equation of the line tangent to the graph of $f(x) = \tan^2 x$ at the point $(\pi/4, 1)$.

$f'(x) = 2\tan x \cdot \sec^2 x$. Evaluated at $\pi/4$, $f'(\pi/4) = 2\tan \pi/4 \cdot \sec^2 \pi/4 = 2(1)(2) = 4$, which is a slope of the tangent line to f(x) at the point $(\pi/4, 1)$. The equation of this tangent line is $y - 1 = 4(x - \pi/4) = 4x - \pi$. Therefore, $y = 4x - \pi + 1$.

Exercises:

1. Find derivative of $(x^3 - 2x^2 + 3x - 1)^7$ [$7(x^3 - 2x^2 + 3x - 1)^6(3x^2 - 4x + 3)$]

2. Find derivative of f(u) with respect to x, $f(u) = u^2 - 6$ and $u(x) = 5x^2 + 1$.

3. Find derivative of g(n) with respect to y, $g(n) = n^3 + n^2 - 5$ and $n(y) = 4y + 6$.

4. Find derivative of $(x^2 + 6x - 2)^{1/2}$ [$(x + 3)(x^2 + 6x - 2)^{-1/2}$]

5. Find derivative of P(m) with respect to w, $P(m) = 2m^2 + 2$ and function $m(w) = 5w^2 + 3w - 3$.

6. Find the derivative of h(a) with respect to b if $h(a) = a^{-4}$ and $u(b) = 6b^3 - b$.

7. Find derivative of $(3x + 2)/(x - 1)$ [$-5/(x - 1)^2$]

8. Find derivative of g(n) with respect to x, $g(n) = n^{1/2}$ and $n(x) = (x^2 - 1)/(x^2 + 2)$.

9. Find derivative of k(q) with respect to x if $k(q) = q^{-3}$ and $q(x) = x^2 - 9$.

10. Given function $r = 0.10p^2$, p is the price, price is dependent on some other variable g given by the formula $p = 0.75g^2 + 15$. Find dr/dg, the derivative of r with respect to g.

IMPLICIT DIFFERENTIATION:

If function is expressed in the form $y = f(x)$, it is said to define y *explicitly* in terms of x. For example, functions $y = 3x^2 - x + 4$, $y = 5(3x^2 + x^7)$, and $y = \sec^3(5x)$ define y explicitly in terms of x. To differentiate such function in one variable x is a simple process of following rules of differentiation:

The derivative of $y = 3x^2 - x + 4$ is $y' = 6x - 1$.
The derivative of $y = 5(3x^2 + x^7)$ is $y' = 5(6x + 7x^6) = 30x + 35x^6 = 5x(6 + 7x^5)$.
The derivative of $y = \sec^3(5x)$ is $y' = 3[\sec(5x)]^2[\sec(5x)\tan(5x)(5)] = 15\sec^3(5x)\tan(5x)$.

Some functions are expressed in two variables. For example, $3x^2 + y - 2 = 0$, $x^2 + y^2 - 25 = 0$, $y - xy^2 + x^2 + 1 = 0$ are functions of both x and y. To differentiate such functions, we must think of the variable y as function of x defined *implicitly*, $y = f(x)$.

$3x^2 + y - 2 = 0$ is $3x^2 + y(x) - 2 = 0$
$x^2 + y^2 - 25 = 0$ is $x^2 + y^2(x) - 25 = 0$
$y - xy^2 + x^2 + 1 = 0$ is $y(x) - xy^2(x) + x^2 + 1 = 0$

The derivative of $3x^2 + y(x) - 2 = 0$ is $(3x^2)' + y' - (2)' = (0)'$, therefore $6x + y' - 0 = 0$ and $y' = -6x$.

The derivative of $x^2 + y^2(x) - 25 = 0$ is $(x^2)' + (y^2)' - (25)' = 0$, which is $2x + 2(y)(y') - 0 = 0$, and therefore, $y' = -(2x)/(2y) = -x/y$.

The derivative of $y - xy^2 + x^2 + 1 = 0$ is $(y)' - (xy^2)' + (x^2)' + (1)' = 0$, which is $y' - [x(2y)(y') + y^2] + 2x = 0$ (we differentiate (xy^2) by Product Rule). Therefore, we obtain $y' - x(2y)(y') - y^2 + 2x = 0$.

Now, to solve for y', we need to get all terms with y' on one side of the equation, that is $y' - x(2y)(y') = y^2 - 2x$. Solving for y', we obtain $(1 - 2xy)y' = y^2 - 2x$ and therefore, $y' = (y^2 - 2x)/(1 - 2xy)$.

Example:

Find derivative of $x^2y^3 + y^5 = 2x + 1$ by using implicit differentiation.

Differentiating implicitly with respect to x, we obtain $2xy^3 + 3x^2y^2y' + 5y^4y' = 2$.
Thus, $y'(3x^2y^2 + 5y^4) = 2 - 2xy^3$ and therefore $y' = (2 - 2xy^3)/(3x^2y^2 + 5y^4)$.

Example:

Find dy/dx of $x^2 + y^2 = y^3 - x$ by implicit differentiation.

Differentiating implicitly, we obtain
$2x + 2yy' = 3y^2y' - 1$
$2yy' - 3y^2y' = -2x - 1$
$(2y - 3y^2)y' = -2x - 1$
Thus, $y' = (-2x - 1)/(2y - 3y^2)$ or $(2x + 1)/(3y^2 - 2y)$.

Example:

Find the slope of the tangent line to the graph of $y^4 + 3y - 4x^3 = 5x + 1$ at $(1, -2)$.

The derivative is $(y^4 + 3y - 4x^3)' = (5x + 1)'$, which is $4y^3y' + 3y' - 12x^2 = 5 + 0$
and so $(4y^3y' + 3y') = 12x^2 + 5$. Therefore, $y'(4y^3 + 3) = 12x^2 + 5$ and
$y' = (12x^2 + 5)/(4y^3 + 3)$.

When $x = 1$ and $y = -2$, the slope is $m = (12(1)^2 + 5)/(4(-2)^3 + 3) = -17/29$.

Example:

Find an equation of a line tangent to the graph of $(y - 3)^2 = 4(x - 5)$ at $(6, 1)$.

Taking the derivatives of both sides, we obtain
$2(y - 3)y' = 4(1)$ and so $y' = 2/(y - 3)$.
At the point $(6, 1)$, the slope is $y' = 2/(1 - 3) = -1$.

An equation of the tangent line is $y - 1 = -1(x - 6)$, which is $y = -x + 7$.

Exercises:

Differentiate the following functions:

1. $x^2 + y^2 = 100$ $[-x/y]$

2. $x^3 - y^3 = 6xy$

3. $3x^2 - xy + 4y^2 = 141$ $[(y - 6x)/(8y - x)]$

4. $x^3y^2 - 5x^2y + x = 1$

5. $x^2y + 3xy^3 - x = 3$ $[(1 - 2xy - 3y^3)/(x^2 + 9xy^2)]$

DIFFERENTIALS:

Given function f(x), any change in x will result in change in y since y = f(x). The change in x is called *increment in x* and change in y is called *increment in y*.

For example, consider function y = f(x) = 2x + 1. If x = 3 then y = 7. If x changes from 3 to 4 then y will change from 7 to 9. The increment in x is 1 and the increment in y is 2. The increment in x is denoted by Δx and the increment in y is denoted by Δy. As we know, the slope of the line (the difference quotient) is defined as "change in y" or Δy, divided by "change in x", or Δx. Thus, the difference quotient can be written as

$\Delta y/\Delta x$ or $\Delta y = f(x + \Delta x) - f(x)$.

The derivative is the limit of the difference quotient as Δx approaches zero. Therefore, we can think of the derivative f(x)' as $\lim_{\Delta x \to 0}(\Delta y/\Delta x)$, if this limit exists. For a sufficiently small (but non-zero) change in x, the difference quotient $\Delta y/\Delta x$ is a good approximation for the derivative f(x)'.
Thus, $\Delta y/\Delta x \approx f(x)'$ and $\Delta y \approx f(x)'\Delta x$.

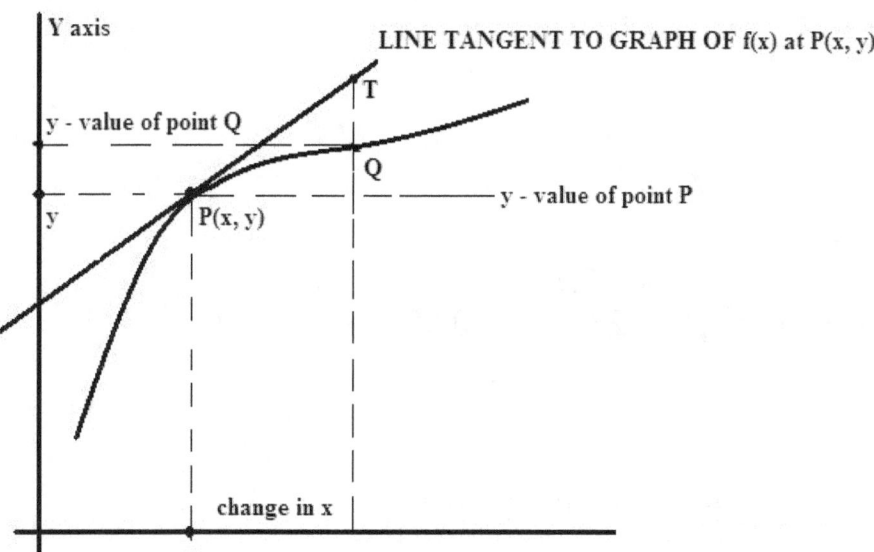

The actual (*not* approximate) change in y is $\Delta y = f(x + \Delta x) - f(x)$. This is simply the difference between y-value of point Q and y-value of point P. The expression f(x)'Δx is an approximation of Δy when change in x is small.

The expression f(x)'Δx is called *differential of y*, denoted by *dy*. The differential of x is denoted by *dx*, where $dx = \Delta x$. Therefore, for a sufficiently small change in x, the definition of the differential of y is $dy = f(x)'dx$, where $dx = \Delta x$.

We can think of *dy* as the distance from point T to the line representing the y-value of point P. When points Q and P are close together, the difference between Δy and *dy* is negligible. That is why $\Delta y \approx dy$.

Example:

Find Δy and *dy* for $f(x) = 1 - 2x^2$ when $x = 0$ and $\Delta x = dx = -0.1$.

$\Delta y = f(x + \Delta x) - f(x) = f(0 - 0.1) - f(0) = (1 - 2(-0.1)^2) - (1 - 2(0)^2) = -0.02$, whereas $dy = f'(x)dx = f'(0)(-0.1) = 0$. Thus, $\Delta y \approx dy$.

Example:

Evaluate *dy* for $f(x) = x^2 + 3x$ when $x = 2$ and $dx = 0.1$.

$dy = f(x)'dx = (2x + 3)dx$. When $x = 2$ and $dx = 0.1$, $dy = (2(2) + 3)(0.1) = 0.7$, whereas $\Delta y = f(x + \Delta x) - f(x) = f(2.1) - f(2) = ((2.1)^2 + 3(2.1)) - (2^2 + 3(2)) =$
$= (4.41 + 6.3) - (4 + 6) = 10.71 - 10 = 0.71$. Thus, $\Delta y \approx dy$.

Example:

Let $y = x^{1/2}$. Find *dy* and Δy if $x = 4$ when $\Delta x = 3$.

$dy = (1/2)x^{-1/2} (dx) = (1/2)(4^{-1/2})(3) = 3/4 = 0.75$.
$\Delta y = (x + \Delta x)^{1/2} - x^{1/2} = (4 + 3)^{1/2} - 4^{1/2} \approx 2.65 - 2 = 0.65$.

Exercises:

1. Find Δy, *dy* for $f(x) = x^3$, $x = 1$, $\Delta x = -0.1$ [$\Delta y = 0.331$, $dy = 0.3$]

2. Find Δy, *dy* for $f(x) = x^4 + 1$ $x = -1$, $\Delta x = -0.01$ [$\Delta y = -0.0394$, $dy = -0.04$]

3. Find Δy, *dy* for $f(x) = x^4 + 1$ $x = 2$, $\Delta x = 0.01$ [$\Delta y = -0.3224$, $dy = -0.32$]

INCREASING AND DECREASING FUNCTIONS:

Function f(x) that is defined on some interval between two numbers x_1 and x_2 may be increasing, decreasing, or constant over this interval.

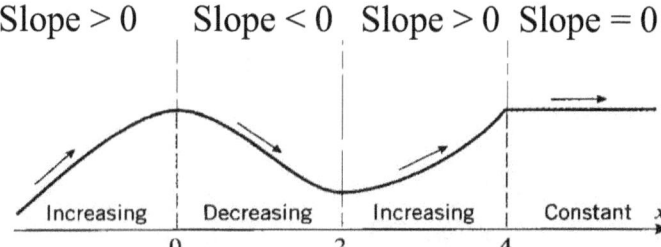

Note: The interval $[x_1, x_2]$ is called "closed interval" when the endpoints x_1 and x_2 are included in the interval as well as all the numbers between x_1 and x_2. The interval (x_1, x_2) is called "open interval" when the endpoints are not included in the interval.

Mean–Value Theorem:

If function f(x) is differentiable on the open interval (a, b) and it is continuous on the closed interval $[a, b]$ then there is a value $x = c$ in (a, b) where $d/dx(f(x)) = (f(b) - f(a))/(b - a)$, which is slope of the line between points O and P.

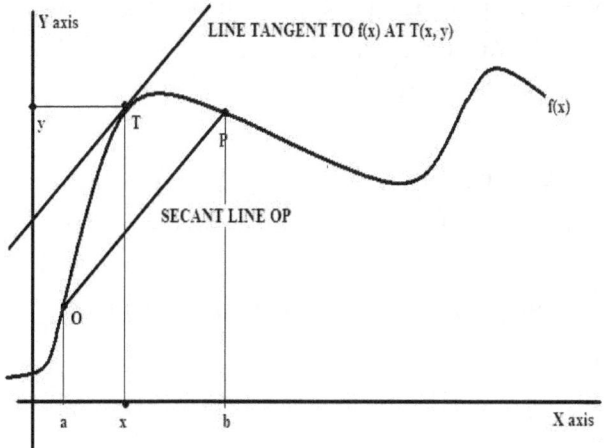

Mean-Value theorem states that if function f(x) is differentiable on (a, b) and continuous on $[a, b]$ then there is at least one point x in (a, b) where the line tangent to the graph of the function f(x) at T(x, y) is parallel to the secant line joining the points O and P. That is, the slope of the secant line OP is the same as the slope of a tangent line evaluated at x.

Example:

Does Mean Value Theorem apply to $f(x) = 4x^3 + 2x + 8$ on the interval $[0, 1]$?

Because f(x) is a polynomial, it is continuous and differentiable everywhere. Therefore, hypotheses of Mean Value theorem are met. According to Mean-Value theorem, $f'(c) = [f(1) - f(0)]/[1 - 0] = 14 - 8/1 = 6$, which is the value of the slope between points where $x = 0$ and $x = 1$.

The derivative $f'(x) = 12x^2 + 2$ must equal to $f'(c) = 6$.
Therefore, $12x^2 + 2 - 6 = 0$ and so $12x^2 - 4 = 0$ and so $4(3x^2 - 1) = 0$.
Thus, $3x^2 - 1 = 0$ and so $x = (1/3)^{1/2}$, which is in the interval $[0, 1]$.

Example:

Show that the function $f(x) = x^3 + x^2$ satisfies the hypotheses of Mean Value Theorem on the interval $[1, 2]$ and find the value $x = c$ such that $f'(c) = [f(2) - f(1)]/(2 - 1)$.

Because f(x) is a polynomial, it is continuous and differentiable everywhere. Therefore hypotheses of Mean Value theorem are met.

The derivative $f'(x) = 3x^2 + 2x$ and $f'(c) = [f(2) - f(1)]/(2 - 1) = 10$.
Therefore, $3x^2 + 2x = 10$ and $3x^2 + 2x - 10 = 0$. Solving the equation using quadratic formula, we obtain $x = (-1 + \sqrt{31})/3$ and so $x = (-1 - \sqrt{31})/3$.
Only the positive value of x is in the interval $[1, 2]$. Thus, $c = (-1 + \sqrt{31})/3$.

This number $c = (-1 + \sqrt{31})/3$ is the x-value where the derivative of f(x) is equal to the slope of the line between the points $(1, f(1))$ and $(2, f(2))$.

Equivalently, the line tangent to function f(x) at point $(c, f(c))$ is parallel to the secant line between points $(1, f(1))$ and $(2, f(2))$.

Rolle's Theorem:

If function f(x) is differentiable on the open interval (a, b) and it is continuous on the closed interval $[a, b]$ where $f(a) = f(b)$ then there is a value $x = c$ in (a, b) such that $d/dx[f(x)] = 0$.

Consider the previous graph between point O where $x = a$ and point P where $x = b$.

If points O and P are on the same "level", meaning that f(a) = f(b), then there exist a point on the graph between x = a and x = b where the tangent line is horizontal.

Consider the previous graph between point O where x = *a* and point P where x = *b*. If point P is "higher" than point O then the function is increasing. Connect points O and P with a line. This secant line points "up", its slope is positive. The theorem tells us that there always exists a point T, whose *x*-value is between *a* and *b*, where the tangent line at T is parallel to the secant line connecting O and P. Thus, the slope of the tangent line is positive and the derivative of f(x) evaluated at x is also positive. Therefore, if function f(x) is differentiable on (x_1, x_2) and continuous on the interval $[x_1, x_2]$:

1. If the derivative of the function is positive for every value of x inside the interval (x_1, x_2) then the function f(x) is increasing on the interval $[x_1, x_2]$.

2. If the derivative of the function is negative for every value of x inside the interval (x_1, x_2) then the function f(x) is decreasing on the interval $[x_1, x_2]$.

3. If the derivative of the function is equal to zero for every value of x inside the interval (x_1, x_2) then the function f(x) is constant (horizontal line) on the interval $[x_1, x_2]$.

To find if the function is increasing or decreasing on the interval:

1. Select any number x in a given interval
2. Substitute this number into the derivative to see if the derivative is positive or negative.

If the derivative is positive then the function is increasing,
If the derivative is negative then the function is decreasing.

Note: function changes from increasing to decreasing or decreasing to increasing at the point x where the derivative of the function is equal to zero. The value of x when the derivative of f(x) is equal to zero is called *critical value, critical point,* or *critical number.* The tangent line to f(x) at this point is a horizontal line (slope zero).

Example:

Given the function $f(x) = x^3 - 3x^2 + 1$, find the intervals on which the function is increasing and decreasing.

First, let's find points for which the derivative of the function is equal to zero.

The derivative of $f(x)$ is $3x^2 - 6x = 3x(x - 2)$
The derivative is zero when $3x(x - 2) = 0$, therefore $x = 0$ or $x = 2$

The critical points $x = 0$ and $x = 2$ divide the number line in three intervals: from $-\infty$ to 0, from 0 to 2, and from 2 to ∞.

Now, we need to select any number in each of the three intervals:
Select $x = -1$ for points less than 0, in the interval $(-\infty, 0)$
Select $x = 1$ for points greater than 0 but less than 2, in the interval $(0, 2)$
Select $x = 5$ for points greater than 2, in the interval $(2, \infty)$.

1. Select any number in the interval $(-\infty, 0)$, for example $x = -1$
2. Substitute $x = -1$ into the derivative $3x(x - 2)$ obtaining $(-3)(-3) = 9$

Since the derivative is 9 (positive), the function is increasing on the interval $(-\infty, 0)$.

1. Select any number in the interval $(0, 2)$, for example $x = 1$
2. Substitute $x = 1$ into the derivative $3x(x - 2)$ obtaining $3(1)(1 - 2) = -3$

Since the derivative is -3 (negative), the function is decreasing on the interval $(0, 2)$.

1. Select any number in the interval $(2, \infty)$, for example $x = 5$
2. Substitute $x = 5$ into the derivative $3x(x - 2)$ obtaining $3(5)(5 - 2) = 45$

Since the derivative is 45 (positive), the function is increasing on the interval $(2, \infty)$.

Example:

Find the intervals on which the function $f(x) = 20x - x^2$ is increasing and decreasing.

To find these intervals we must find values of x for which the derivative of the function is equal to zero. The derivative of $f(x) = 20x - x^2$ is $20 - 2x$. The derivative is zero when $20 - 2x = 0$, therefore $20 = 2x$, and $x = 10$. Point $x = 10$ (critical number) divides the number line into two intervals:

all points less than 10, i.e, negative infinity to 10 $(-\infty, 10)$ and
all points greater than 10, i.e, from 10 to positive infinity $(10, \infty)$

1. Select any number in the interval $(-\infty, 10)$, for example $x = -4$
2. Substitute $x = -4$ into the derivative $20 - 2x$, obtaining $20 - 2(-4) = 2$

Since the derivative is 2 (positive), the function is increasing on the interval $(-\infty, 10)$.

1. Select any number in the interval $(10, \infty)$, for example $x = 13$
2. Substitute $x = 13$ into the derivative $20 - 2x$, obtaining $20 - 2(13) = -6$

Since the derivative is -6 (negative), the function is decreasing on the interval $(10, \infty)$.

Example:

Given $f(x) = 2x^3 + 3x^2 - 12x + 1$, find the intervals on which the function is increasing and decreasing.

To find these intervals we must find values of x for which the derivative of the function is equal to zero.

The derivative of $f(x)$ is $6x^2 + 6x - 12 = 6(x^2 + x - 2) = 6(x + 2)(x - 1)$.

The derivative is zero when $6(x + 2)(x - 1) = 0$, therefore $(x + 2) = 0$ or $(x - 1) = 0$, therefore $x = -2$ or $x = 1$. Thus, the intervals are: $(-\infty, -2)$, $(-2, 1)$, and $(1, \infty)$.

1. Select any number in each interval
2. Substitute this number into the derivative to see if the derivative is positive or negative:

1. Select any number in the interval $(-\infty, -2)$, for example $x = -3$
2. Substitute $x = -3$ into $6x^2 + 6x - 12$, obtaining $6(-3)^2 + 6(-3) - 12 = 24$

Since the derivative is 24 (positive), the function is increasing on the interval $(-\infty, -2)$.

1. Select any number in the interval $(-2, 1)$, for example $x = 0$
2. Substitute $x = 0$ into $6x^2 + 6x - 12$, obtaining $6(0)^2 + 6(0) - 12 = -12$

Since the derivative is -12 (negative), the function is decreasing on the interval $(-2, 1)$.

1. Select any number in the interval $(1, \infty)$, for example $x = 3$
2. Substitute $x = 3$ into $6x^2 + 6x - 12$, obtaining $6(3)^2 + 6(3) - 12 = 60$

Since the derivative is 60 (positive), the function is increasing on the interval $(1, \infty)$.

Exercises:

Find the intervals on which the following functions are increasing and decreasing:

1. $f(x) = x^2 - 2x - 8$ [increasing on $(1, \infty)$, decreasing on $(-\infty, 1)$]

2. $f(x) = x(x - 3)$

3. $f(x) = x^2 - 2x$

4. $f(x) = 27x - x^3$ [increasing on $(-3, 3)$, decreasing on $(-\infty, -3)$ and $(3, \infty)$]

5. $f(x) = x^2 - 5x + 6$

6. $f(x) = 3x^2 - 4x + 3$

7. $f(x) = (x + 2)^2(x - 1)$ [increasing on $(-\infty, -2)$, $(0, \infty)$, decreasing on $(-2, 0)$]

FIRST DERIVATIVE TEST:

The point at which a continuous function changes from increasing to decreasing is called *relative maximum*. It is the 'highest' point on a graph over the interval (x_1, x_2). The point at which a continuous function changes from decreasing to increasing is called *relative minimum*. It is the 'lowest' point on a graph over the interval (x_1, x_2). These points are called *relative extrema* or *local extrema*. Given function f(x) defined over some interval (x_1, x_2) we can find all relative extrema in this interval by following these steps:

1. Compute the derivative of f(x).

2. Set the derivative equal to zero and solve for x. The result is critical value(s), where the function changes from increasing to decreasing or decreasing to increasing.

3. These critical values divide number into intervals. Select one number in each interval and substitute these chosen numbers into the derivative.

If the derivative is negative, the function is decreasing.
If the derivative is positive, the function is increasing.

4. Identify value or values of x that are relative extrema: when the sign of the derivative changes from plus to minus, we have relative maximum. When the sign of the derivative changes from minus to plus, we have relative minimum.

5. Compute y values by substituting critical values from step 2 into the original function f(x).

Example:

Find all extrema for the function $f(x) = x^3 - 3x^2 + 1$.

The derivative of f(x) is $3x^2 - 6x = 3x(x - 2)$
The derivative is zero when $3x(x - 2) = 0$, therefore $x = 0$ or $x = 2$ (critical values).
Thus, the intervals are $(-\infty, 0), (0, 2), (2, \infty)$.

Select $x = -1$ for points less than 0 in $(-\infty, 0)$
Select $x = 1$ for points greater than 0 but less than 2 in $(0, 2)$
Select $x = 5$ for points greater than 2 in $(2, \infty)$

1. Select any number in the interval $(-\infty, 0)$, for example $x = -1$
2. Substitute $x = -1$ into the derivative $3x(x-2)$ obtaining $(-3)(-3) = 9$

Since the derivative is 9 (positive), the function is increasing on the interval $(-\infty, 0)$.

1. Select any number in the interval $(0, 2)$, for example $x = 1$
2. Substitute $x = 1$ into the derivative $3x(x-2)$ obtaining $3(1)(1-2) = -3$

Since the derivative is -3 (negative), the function is decreasing on the interval $(0, 2)$.

1. Select any number in the interval $(2, \infty)$, for example $x = 5$
2. Substitute $x = 3$ into the derivative $3x(x-2)$ obtaining $3(5)(5-2) = 45$

Since the derivative is 45 (positive), the function is increasing on the interval $(2, \infty)$.

Therefore, the graph of $f(x) = f(x) = x^3 - 3x^2 + 1$ is

increasing to the left of $x = 0$, on the interval $(-\infty, 0)$
decreasing from $x = 0$ to $x = 2$, on the interval $(0, 2)$
increasing to the right of $x = 2$, on the interval $(2, +\infty)$

Thus, the maximum occurs at $x = 0$ and the minimum occurs at $x = 2$
Substituting the critical values into the original function $f(x) = x^3 - 3x^2 + 1$, we obtain

$f(0) = (0)^3 - 3(0)^2 + 1 = 1$, the point $(0, 1)$ is maximum
$f(2) = (2)^3 - 3(2)^2 + 1 = -3$, the point $(2, -3)$ is minimum.

Example:

Find all local of $f(x) = x^3 - 3x - 1$.

1. The derivative of $f(x)$ is $3x^2 - 3$
2. Setting the derivative equal to zero we obtain
$3x^2 - 3 = 0$
$3(x^2 - 1) = 0$

$x^2 - 1 = 0$
$x^2 = 1$
$x = 1$ or -1 (alternatively, we can factor $x^2 - 1$ into $(x - 1)(x + 1) = 0$, thus $x - 1 = 0$ or $x + 1 = 0$ which gives us the same result: $x = 1$ or $x = -1$). Therefore, the critical values are $x = 1$ and $x = -1$. These values divide the number line into three intervals $(-\infty, -1)$, $(-1, 1)$, and $(1, \infty)$.

3. Select -2 because it is less than the smallest critical value -1, select 0 because it is between critical values -1 and 1, and select 2 because it is greater than the largest critical value 1.

Substitute these numbers into the derivative. If the derivative is negative, function is decreasing. If the derivative is positive, function is increasing:

If $x = -2$, $3x^2 - 3 = 3(-2)^2 - 3 = 9$, function is increasing
If $x = 0$, $3x^2 - 3 = 3(0)^2 - 3 = -3$, function is decreasing
If $x = 2$, $3(2)^2 - 3 = 9$, function is increasing

4. Identify values that are extrema:

The sign of the derivative changes from plus to minus at $x = -1$, therefore it is maximum. The sign of the derivative changes from minus to plus at $x = 1$, therefore it is minimum.

5. Compute the values of y corresponding to the values of x of extrema by substituting critical values $x = 1$ and $x = -1$ into the original function $f(x) = x^3 - 3x - 1$:

$f(-1) = (-1)^3 - 3(-1) - 1 = 1$, then $y = 1$, $x = -1$, thus point $(-1, 1)$ is maximum. Since $f(1) = (1)^3 - 3(1) - 1 = -3$, $y = -3$, $x = 1$, the point $(1, -3)$ is minimum.

Example:

Find all extrema of $f(x) = 2x^3 - 3x^2 - 12x + 1$.

The derivative of $f(x)$ is $6x^2 - 6x - 12 = 6(x^2 - x - 2) = 6(x + 1)(x - 2)$. Setting the derivative equal to zero we obtain $x + 1 = 0$ or $x - 2 = 0$. Therefore, the critical values are $x = -1$ and $x = 2$. The corresponding intervals are $(-\infty, -1)$, $(-1, 2)$, and $(2, \infty)$.

Select x = − 2 in (− ∞, − 1)
Select x = 0 in (− 1, 2)
Select x = 3 in (2, ∞)

For x = − 2, the derivative $6(x + 1)(x - 2)$ is equal to 24 (positive)
For x = 0, the derivative $6(x + 1)(x - 2)$ is equal to − 12 (negative)
For x = 3, the derivative $6(x + 1)(x - 2)$ is equal to 24 (positive)

Therefore, the function $f(x) = 2x^3 - 3x^2 - 12x + 1$ increases on (− ∞, − 1), decreases on (− 1, 2), and increases on (2, ∞).

The value of the function $f(x) = 2x^3 - 3x^2 - 12x + 1$ at the critical value x = − 1 is $2(-1)^3 - 3(-1)^2 - 12(-1) + 1 = 8$, therefore the maximum point is (− 1, 8).

The value of the function $f(x) = 2x^3 - 3x^2 - 12x + 1$ at the critical value x = 2 is $2(2)^3 - 3(2)^2 - 12(2) + 1 = -19$, therefore the minimum point is (2, − 19).

Exercises:

Use First Derivative Test to find local extrema (min/max):

1. 7. $f(x) = (x + 2)^2(x - 1)$ [maximum (− 2, 0), minimum (0, − 4)]

2. $f(x) = x^4 - 2x^2 + 3$

3. $f(x) = x^3 + 3x^2 + 3x - 4$

4. $f(x) = x^4 - 32x + 4$ [minimum (2, − 44)]

5. $f(x) = (x + 1)^6$

9. $f(x) = (x + 2)^{1/3}$

10. $f(x) = (x)^{1/3}(x + 4)$

SECOND DERIVATIVE TEST:

It is usually simpler to use Second Derivative Test to find extrema (min/max):

1. Compute the derivative of f(x).
2. Set the derivative equal to zero and solve for x. The solutions are critical values.
3. Compute the second derivative of f(x).
4. Substitute critical values from Step 2 into the second derivative.

If the second derivative is positive, then function f(x) has minimum at that point.
If the second derivative is negative, then function f(x) has maximum at that point.

Note: If the second derivative is zero then the test fails. In that case use First Derivative Test.

Example:

Determine the points where the graph of $f(x) = x^4 - 2x^2$ has maximum and minimum.

The first derivative of f(x) is $4x^3 - 4x = 4x(x - 1)(x + 1)$. Set derivative equal to zero and solve for x: $4x(x - 1)(x + 1) = 0$, therefore $x = 0$, $x = 1$, and $x = -1$ are critical values.

The second derivative is $12x^2 - 4$ (it is the derivative of $4x^3 - 4x$). Substituting critical values into the second derivative we obtain

$12(0)^2 - 4 = -4$ which is negative (maximum)
$12(1)^2 - 4 = 8$ which is positive (minimum)
$12(-1)^2 - 4 = 8$ which is positive (minimum)

To be able to plot these points (extrema) on the graph, we need the values of y corresponding to critical values $x = -1, 0, 1$. We obtain the values of the function by substituting critical values into the original function

$y = f(x) = x^4 - 2x^2$:
$y = f(-1) = (-1)^4 - 2(-1)^2 = 1 - 2 = -1$, point $(-1, -1)$ is minimum
$y = f(0) = (0)^4 - 2(0)^2 = 0$, point $(0, 0)$ is maximum
$y = f(1) = (1)^4 - 2(1)^2 = -1$, point $(1, -1)$ is minimum.

Example:

Determine the points where the graph of $f(x) = 8x^3 - 3x^4$ has maximum and minimum.

The first derivative is $24x^2 - 12x^3$. Set derivative equal to zero and solve for x: $24x^2 - 12x^3 = 0$. Factor $12x^2$ out of $24x^2$ and $12x^3$ obtaining $12x^2(2 - x) = 0$, therefore $12x^2 = 0$ or $2 - x = 0$, thus $x = 0$ or $x = 2$ are critical values.

The second derivative of f(x) is $d/dx[24x^2 - 12x^3] = 48x - 36x^2$

Evaluating second derivative at critical value $x = 0$ we obtain $48(0) - 36(0)^2 = 0$
Evaluating second derivative at critical value $x = 2$ we obtain $48(2) - 36(2)^2 = -48$

Since the second derivative at $x = 0$ is zero, the test fails. The function may or may not have minimum or maximum at this point.

The second derivative evaluated at another critical value $x = 2$ is -48. Thus, the function has maximum at $x = 2$. What is the value of $y = f(2) = 8x^3 - 3x^4$?

The value of $y = f(2) = 8(2)^3 - 3(2)^4 = 8(8) - 3(16) = 64 - 48 = 16$. Therefore, the maximum occurs at the point (2, 16).

Example:

Find extrema of the function $f(x) = 2x^2 - x^4$.

The derivative is $4x - 4x^3 = 4x(1 - x^2)$.
The derivative $4x(1 - x^2)$ is zero when $x = 0$, $x = 1$, or $x = -1$.

These are critical values. The graph of f(x) has a horizontal tangent at point(s) where derivative of f(x) is equal to zero.

The second derivative of f(x) is $4 - 12x^2$. Evaluating second derivative at critical value $x = 0$ we obtain $4 - 12(0)^2 = 4$, which is positive. Evaluating second derivative at critical value $x = 1$ we obtain $4 - 12(1)^2 = -8$, which is negative. Evaluating second derivative at critical value $x = -1$ we obtain $4 - 12(-1)^2 = -8$, which is also negative. Therefore, minimum is at $x = 0$ and maximum is at $x = 1$ and $x = -1$.

When x = 0, y = f(x) = 2(0)² − (0)⁴ = 0, the point (0, 0) is minimum,
When x = 1, y = f(x) = 2(1)² − (1)⁴ = 1, the point (1, 1) is maximum,
When x = − 1, y = f(x) = 2(− 1)² − (− 1)⁴ = 1, the point (− 1, 1) is also maximum.

Exercises:

Use Second Derivative Test to find extrema and the corresponding values of f(x) for the following functions:

1. $f(x) = (x - 5)^2$ [minimum (5, 0)]

2. $f(x) = x^4 - 2x^2 + 3$

3. $f(x) = x^3 + 3x^2 + 3x - 4$

4. $f(x) = x^3 - 3x^2 + 3$ [minimum (2, − 1), maximum (0, 3)]

5. $f(x) = (x + 1)^4$

6. $f(x) = x^3 - 3x^2$

7. $f(x) = (1/3)x^2 - 3$ [if test fails, use 1ˢᵗ Derivative test, minimum (0, − 3)]

8. $f(x) = 2(x + 1)^2 - 8(x + 20) + 7$

9. $f(x) = (x^3 + 3x^2)^3$

10. $f(x) = (x + 2)^{1/3}$

CONCAVITY AND INFLECTION POINTS:

Where the graph of f(x) is "curved up" we say that the function f(x) is *concave up*; where the graph is "curved down" we say that the function f(x) is *concave down*.

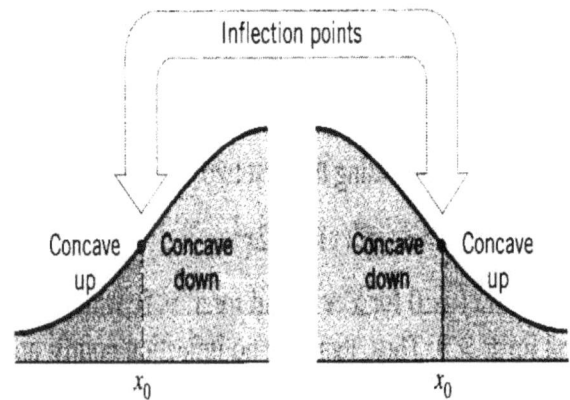

The Second Derivative Test is used to determine concavity:

If the second derivative of the function f(x) is positive on a given interval, then the function is *concave up* on that interval.

If the second derivative of the function f(x) is negative on a given interval, then the function is *concave down* on that interval.

To find the points where concavity may be changing, we need to find the values of x for which the second derivative is zero or second derivative does not exist.

Then we need to determine the sign of the second derivative in each of the test intervals.

If there are any x-values for which the function f(x) is itself undefined, then concavity may be changing at these points as well.

The reason is that if the second derivative of f(x) is negative, then the first derivative decreases as x increases and so the tangent line turns clockwise. Whereas, if the second derivative of f(x) is positive, then the first derivative increases as x increases and so the tangent line turns counterclockwise.

Example:

Find the interval at which $f(x) = x^2 - 4x + 3$ is concave up and the interval at which it is concave down.

The derivative of f(x) is $2x - 4$. The second derivative is 2.
The second derivative is always positive, therefore the function $x^2 - 4x + 3$ is concave up on the interval $(-\infty, +\infty)$. The graph of $f(x) = x^2 - 4x + 3$ is a parabola.

Example:

Find the interval at which $f(x) = x^3 - 3x^2 + 1$ is concave up and the interval at which it is concave down.

The derivative of f(x) is $3x^2 - 6x$. The second derivative is $6x - 6 = 6(x - 1)$. To find out when the second derivative is positive or negative, we need to see when it is equal to zero. The second derivative is equal to zero when $x = 1$. Therefore, the second derivative is positive when $x > 1$, the second derivative is negative when $x < 1$. Thus, the function $f(x) = x^3 - 3x^2 + 1$ is concave up on the interval $(1, +\infty)$ and concave down on the interval $(-\infty, 1)$.

Example:

Find the interval at which $f(x) = x^3$ is concave up and the interval at which it is concave down.

The derivative of f(x) is $3x^2$ and the second derivative is $6x$.
The second derivative $6x$ is positive if x is positive and negative if x is negative. Thus the function $f(x) = x^3$ is concave down to the left of the y–axis (where x is negative) and concave up to the right of y-axis (where x is positive): f(x) is concave down on the interval $(-\infty, 0)$ and concave up on the interval $(0, +\infty)$.

Example:

Find the interval at which $f(x) = x^4 - 6x^2 + 2$ is concave up and the interval at which it is concave down.

The first derivative of f(x) is $4x^3 - 12x$ and the second derivative is $12x^2 - 12$. To find out when the second derivative is positive or negative, we need to see when it is equal to zero. Second derivative is equal to zero when $12x^2 - 12 = 0$, $12x^2 = 12$, $x^2 = 1$, and therefore $x = 1$ or $x = -1$, dividing the number line into three intervals: $(-\infty, -1)$, $(-1, 1)$ and $(1, +\infty)$.

We need to determine the sign of the second derivative on each of these intervals by choosing a number in each interval and substituting this number into the second derivative $12x^2 - 12$.

Let's select $x = -2$ in $(-\infty, -1)$, $x = 0$ in $(-1, 1)$ and $x = 2$ in $(1, +\infty)$. Then the second derivative $12x^2 - 12$ is

$12(-2)^2 - 12 = 36$ positive on $(-\infty, -1)$
$12(0)^2 - 12 = -12$ negative on $(-1, 1)$
$12(2)^2 - 12 = 36$ positive on $(1, +\infty)$

Thus, the graph of $f(x) = x^4 - 6x^2 + 2$ is concave up on $(-\infty, -1)$, concave down on $(-1, 1)$, and concave up on $(1, +\infty)$.

The concavity of the graph changes at the points where the second derivative is zero, which is at $x = -1$ and $x = 1$. The values of the function $f(x) = x^4 - 6x^2 + 2$ at these points are: $f(-1) = (-1)^4 - 6(-1)^2 + 2 = -3$ at the point $(-1, -3)$, and $f(1) = (1)^4 - 6(1)^2 + 2 = -3$ at the point $(1, -3)$. These points are called *inflection points*.

Inflection Points:

The point at which function changes its concavity is called an *inflection point*. Inflection points mark the places on the graph of f(x) where the derivative of f(x) changes from increasing to decreasing, or decreasing to increasing.

If the point (c, f(c)) is an inflection point then the second derivative of f(x) is either zero or undefined at $x = c$, *and* the second derivative of f(x) changes sign at this point (that is, concavity changes at $x = c$). The inflection point is a point on a graph where the second derivative is positive for the values of x on one side of the point and negative on the other.

Note: it is possible that the second derivative may be equal to zero at some point P that is not an inflection point. If concavity does not change at the point P, it is not an inflection point even though the second derivative at this point may be zero.

Example:

Find intervals of concavity and inflection point(s) of $f(x) = x^4 - 8x^3 + 12x - 5$.

The derivative of $f(x)$ is $4x^3 - 24x^2 + 12$
The second derivative is $12x^2 - 48x = 12x(x - 4)$

Second derivative equal to zero means $12x(x - 4) = 0$ thus $12x = 0$ or $x - 4 = 0$, therefore $x = 0$ or $x = 4$. The values $x = 0$ and $x = 4$ divide the number line into three intervals: $(-\infty, 0)$, $(0, 4)$, and $(4, +\infty)$. We need to determine the sign of the second derivative in each of these intervals by arbitrarily choosing a number in each interval and substituting this number into the second derivative $12x^2 - 48x$.

Let's select $x = -1$ in $(-\infty, 0)$, $x = 1$ in $(0, 4)$, and $x = 6$ in $(4, +\infty)$.
Then the second derivative $12x^2 - 48x$ evaluated at these points is equal to

$12(-1)^2 - 48(-1) = 60$ positive on $(-\infty, 0)$
$12(1)^2 - 48(1) = -36$ negative on $(0, 4)$
$12(6)^2 - 48(6) = 144$ positive on $(4, +\infty)$

Therefore, the graph of the function $f(x) = x^4 - 8x^3 + 12x - 5$ is
concave up on $(-\infty, 0)$
concave down on $(0, 4)$
concave up on $(4, \infty)$

The concavity of the graph changes at the points where the second derivative is zero at $x = 0$ and $x = 4$. The values of the function $f(x) = x^4 - 8x^3 + 12x - 5$ at these point are

$f(0) = (0)^4 - 8(0)^3 + 12(0) - 5 = -5$
$f(4) = (4)^4 - 8(4)^3 + 12(4) - 5 = -213$
The inflection points are $(0, -5)$ and $(4, -213)$.

Example:

Find inflection point(s) of $f(x) = x^3$.

The derivative of $f(x) = x^3$ is $3x^2$ and the second derivative is $6x$.
The second derivative is equal to zero when $x = 0$. Thus a possible inflection point

of $f(x) = x^3$ is at $x = 0$. But is it an actual inflection point? In other words, does the concavity of $f(x)$ changes at $x = 0$? The second derivative $6x$ is positive when x is positive and it is negative when x is negative. Since concavity changes at $x = 0$, it is an actual inflection point.

Note: It is possible for the second derivative to be zero at some point x that is not an inflection point. For example, when second derivative is never negative because the function is always concave up, like it is for $f(x) = x^2$.

Example:

Find inflection point(s) of $f(x) = x^4$.

The derivative of $f(x)$ is $4x^3$ and the second derivative is $12x^2$.
The second derivative is zero when $x = 0$.

However, the second derivative does not change the sign at $x = 0$ because $12x^2$ is never negative. The graph of $f(x) = x^4$ is always concave up.

Example:

Find inflection point(s) of $f(x) = x^{1/3}$.

The derivative of $f(x)$ is $(1/3)x^{-2/3}$. Critical value is $x = 0$.
The second derivative is $(1/3)(-2/3)x^{-2/3-1} = -(2/9)x^{-5/3} = -2/(9x^{5/3})$
The second derivative is undefined when $x = 0$. The point $x = 0$ is the inflection point even though the second derivative does not exist at $x = 0$ because concavity changes at $x = 0$. The graph of the function is concave up for negative x and concave down for positive x.

Example:

Find inflection points of $f(x) = -x^2$.

The first derivative of $f(x)$ is $-2x$ and the second derivative of $f(x)$ is -2. The second derivative is always negative. The function $f(x) = -x^2$ is concave down for all x. This function is simply an inverted parabola.

Example:

Does the function $y = x^{1/3} + 2$ have any inflection points? If so, what are they?

The derivative of y is $(1/3)x^{-2/3}$ and the second derivative is $(-2/9)x^{-5/3} = -2/9x^{5/3}$. The second derivative is never zero, but it does not to exist at $x = 0$, which is a vertical asymptote. But $x = 0$ is in the domain of the original function so we can test the intervals $(-\infty, 0)$ and $(0, +\infty)$. The second derivative is greater than zero for $x < 0$ (concave up) while it is less than zero for $x > 0$ (concave down). Since concavity changes at $x = 0$, we have an actual inflection point at $(0, f(0)) = (0, 2)$.

Exercises:

Use Second Derivative Test to determine concavity and find inflection points for the following functions:

1. $f(x) = -(x-5)^2$ [concave down for all x, maximum (5, 0)]

2. $f(x) = (x-1)^3$

3. $f(x) = -3x^5 + 5x^3$

4. $f(x) = x/(x-1)$ [concave down for x < 1, concave up for x > 1, no inflection]

5. $f(x) = x^3 - 9x^2 + 7x$

6. $f(x) = x^3 - 12x$

7. $f(x) = (x^2 + 1)^{1/2}$ [concave up for all x, minimum (0, 1)]

8. $f(x) = (2x - 1)^5$

9. $f(x) = (x^3 + 3x^2)^3$

10. $f(x) = (x + 2)^{1/3}$

ABSOLUTE MAXIMUM AND ABSOLUTE MINIMUM:

Function f(x) has absolute maximum/minimum over the interval $[x_1, x_2]$ at point (x, y) when the value of y = f(x) is the largest/smallest value of the function on this interval $[x_1, x_2]$.

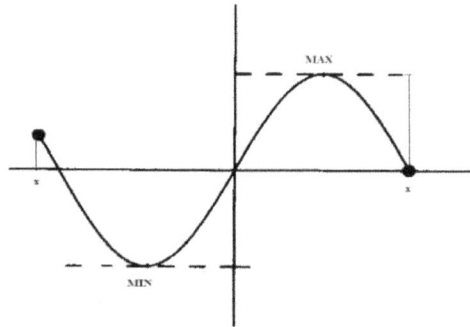

Extreme–Value Theorem:

If function is continuous on a closed interval then it has both absolute maximum and minimum (called *absolute extrema*) on that interval. To find the absolute extrema:

1. Find critical values of the function by setting the derivative of the function f(x) equal to zero and solving for x.

2. Evaluate f(x) at the critical values *and* at the endpoints x_1, x_2 of the interval $[x_1, x_2]$.

The largest value of the function is the absolute maximum and the smallest value of the function is the absolute minimum on the interval $[x_1, x_2]$.

Example:

Find all maximum and minimum of the function $f(x) = x^3 - 3x + 2$.

The domain has no endpoints and the function is differentiable everywhere, therefore *extrema* occur only where 1st derivative of the function is zero:
$f(x)' = 3x^2 - 3 = 3(x - 1)(x + 1)$. If $3(x - 1)(x + 1) = 0$ therefore, critical values are x = 1 and x = −1. Evaluating the function $f(x) = x^3 - 3x + 2$ at these critical values we obtain

$f(-1) = (-1)^3 - 3(-1) + 2 = 4$, therefore $(-1, 4)$ is maximum
$f(1) = (1)^3 - 3(1) + 2 = 0$, therefore $(1, 0)$ is minimum

Using Second Derivative Test, we obtain the same results. The second derivative of $f(x)$ is $6x$. The second derivative is negative at $x = -1$ and positive at $x = 1$. Therefore, $(-1, 4)$ is maximum and $(1, 0)$ is minimum.

Example:

Find all maxima and minima of $f(x) = x^3 - 3x^2 - 24x + 5$ on the interval $[-3, 8]$.

The derivative of $f(x)$ is $3x^2 - 6x - 24 = (3x + 6)(x - 4)$.
The derivative is zero if $3x + 6 = 0$, $x = -2$ or $x - 4 = 0$, $x = 4$.

Thus $x = -2$ and $x = 4$ are critical values if the function. To find all maxima and minima on the interval $[-3, 8]$ we need to evaluate $f(x) = x^3 - 3x^2 - 24x + 5$ at critical values -2 and 4, and the endpoints -3 and 8:

$f(-2) = (-2)^3 - 3(-2)^2 - 24(-2) + 5 = 33$
$f(4) = (4)^3 - 3(4)^2 - 24(4) + 5 = -75$
$f(-3) = (-3)^3 - 3(-3)^2 - 24(-3) + 5 = 23$
$f(8) = (8)^3 - 3(8)^2 - 24(8) + 5 = 133$

The maximum on the interval $[-3, 8]$ is 133 and the minimum is -75.

Example:

Find two positive numbers whose sum is 20 and whose product is as large as possible.

If one number is x then another is $(20 - x)$ and the product is $x(20 - x)$. This is the function that we need to maximize: $f(x) = x(20 - x) = 20x - x^2$. This function is differentiable everywhere, x is between 0 and 20. The maximum is located either at the endpoints 0 or 20, or at a point x where the derivative of $f(x)$ is zero. The derivative of $f(x)$ is $20 - 2x = 2(10 - x)$. The derivative is zero when $x = 10$.

The absolute maximum is at one of the following three points: endpoints $x = 0$, $x = 20$, or the point at which derivative of the function is zero, which is the point where $x = 10$.

Evaluating $f(x) = 20x - x^2$ at these points, we obtain:

$f(0) = 20x - x^2 = 20(0) - (0)^2 = 0$
$f(20) = 20x - x^2 = 20(20) - (20)^2 = 0$
$f(10) = 20x - x^2 = 20(10) - (10)^2 = 100.$

Thus, the maximum of function $f(x) = 20x - x^2$ is 100 when x is 10, and so the two positive numbers that maximize the area are $x = 10$ and $x = 20 - 10 = 10$.

Example:

Find the dimensions of a rectangle with perimeter 100 ft whose area is as large as possible.

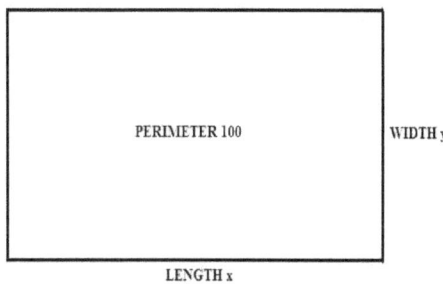

x = length
y = width
A = area

The area $A = xy$ and the perimeter $P = 2x + 2y = 100$. We must maximize the perimeter function and so must express it as function y of x: $2x + 2y = 100$. Therefore, $2y = 100 - 2x$ and so $y = 50 - x$.
Then area $A = xy = x(50 - x) = 50x - x^2$.

Since the two sides of length x cannot be longer than the perimeter $P = 100$, x must be between 0 and 50 ft. Thus, we need to maximize this function $50x - x^2$ over the interval $[0, 50]$. The derivative of $50x - x^2$ is $50 - 2x$. Setting the derivative equal to zero and solving for x, we obtain

$50 - 2x = 0$
$50 = 2x$
$x = 25$

Thus, the maximum occurs at $x = 0$, $x = 25$ or $x = 50$. Substituting these values into the area formula $A(x) = 50x - x^2$, we obtain

$A(0) = 50(0) - (0)^2 = = 0$
$A(25) = 50(25) - (25)^2 = 625$
$A(50) = 50(50) - (50)^2 = 0$.

The maximum area is 625 ft when the sides are 25 ft each.

Example:

Find the minimum length of fencing needed to fence 5000 square feet parking along the road. The parking must be fenced on three sides only.

Let $f(x)$ denote the length of fencing. Then $f(x) = x + 2y$ (length plus two widths). Since the total area is 5000 feet, $xy = 5000$ and $y = 5000/x$.

The length of fencing $f(x)$ can be written in terms of only one variable x. Substituting $y = 5000/x$ into the equation $f(x) = x + 2y$.
Thus, $f(x) = x + 2(5000/x) = x + 10000/x$.

The length of fencing can only be positive, so we must minimize function $f(x)$ on the interval $(0, \infty)$.

The length of fencing function $f(x) = x + 10000/x = x + 10000(x^{-1})$
The derivative of this function is $1 - 10000(x^{-2}) = 1 - 10000/x^2$

Setting the derivative equal to zero we obtain $1 - 10000/x^2 = 0$, thus $10000/x^2 = 1$ and $x^2 = 10000$, thus $x = 100$ or $x = -100$. Since the length of fencing can only be positive, we have two intervals $(0, 100)$ and $(100, \infty)$. The derivative $1 - 10000/x^2$ is negative for x in $(0, 100)$ and positive for x in $(100, \infty)$. The value of $x = 100$ is the point where the derivative is zero - the minimum point. The smallest amount of fencing is $f(100) = 100 + 10000/100 = 200$ feet.

Example:

In a clothing store the amount of sales transactions seems to change with the time of day. The amount is given by the formula $A(x) = 2x^3 - 21x^2 + 60x + 40$ where x is time from 1pm to 6pm. Determine the time of the greatest and the lowest sale.

We need to find maximum and minimum of the function
$A(x) = 2x^3 - 21x^2 + 60x + 40$ on the interval $[1, 6]$.

The derivative of this function is $6x^2 - 42x + 60 = 6(x^2 - 7x^2 + 10)$
The derivative is zero when $(x^2 - 7x^2 + 10) = (x - 2)(x - 5) = 0$

Therefore $x = 2$ and $x = 5$ are the critical values. Since we are only interested in the interval from 1 pm to 6 pm, we need to find the value of the function $A(x)$ at the endpoints 1 and 6, and the critical values 2 and 5:

$A(1) = 2(1)^3 - 21(1)^2 + 60(1) + 40 = 81$
$A(2) = 2(2)^3 - 21(2)^2 + 60(2) + 40 = 92$
$A(5) = 2(5)^3 - 21(5)^2 + 60(5) + 40 = 65$
$A(6) = 2(6)^3 - 21(6)^2 + 60(6) + 40 = 76$

The largest sales transaction ($92) is made at 2 pm and the smallest ($65) is made at 5 pm.

Example:

For $f(x) = (x - 1)/(x - 2)$, find

a. all asymptotes
b. intervals of increase and decrease
c. Minimum and Maximum points
d. intervals on which $f(x)$ is concave up or down
e. inflection point(s)

a. Function $f(x)$ is undefined for $x = 2$ (vertical asymptote at $x = 2$) and
$\lim_{x \to \infty} (x - 1)/(x - 2) = 1/1 = 1$ (horizontal asymptote).

b. Derivative $f'(x) = [(x - 2)(1) - (x - 1)(1)]/(x - 2)^2 = - 1/(x - 2)^2 < 0$ for all x, so $f(x)$ is decreasing for all x.
c. No Minimum or Maximum.

d. Second derivative $f''(x) = 2/(x - 2)^3$ is never zero and $f''(x)$ is undefined for $x = 2$ (possible inflection point). Second derivative $f''(x) < 0$ on $(-\infty, 2)$, where function is concave down and $f''(x) > 0$ on $(2, \infty)$, where function is concave up.

e. Therefore, $x = 2$ is inflection point.

Example:

For $f(x) = (x^2 - x - 2)/(x - 3)$, find

a. all asymptotes
b. intervals of increase and decrease
c. Minimum and Maximum points
d. intervals on which f(x) is concave up or down and
e. inflection point(s)

a. Function $f(x) = (x^2 - x - 2)/(x - 3) = x + 2 + (4/(x - 3))$ so the equation for the slant asymptote is $y = x + 2$. No horizontal asymptote since $\lim_{x \to \infty} f(x) = \infty$. Function is undefined at $x = 3$ since $(x - 3) = 0$ when $x = 3$ (vertical asymptote).

b. Derivative $f'(x) = (x^2 - 6x + 5)/(x - 3)^2$, thus critical values are $x = 5$ or $x = 1$. Second derivative $f''(x) = 8/(x - 3)^{-3}$ thus $f''(x) \neq 0$ for all real numbers and $f''(x)$ is undefined for $x = 3$ (possible inflection point). Using 1st Derivative Test and testing the intervals between critical values 5 and 1, we see that f(x) is increasing on $(-\infty, 1)$, decreasing on $(1, 5)$, and increasing on $(5, \infty)$.

c. Using 2nd Derivative Test, we see that $f''(5) > 0$, therefore, minimum is at $x = 5$, whereas $f''(1) < 0$, therefore, maximum at $x = 1$.

d. Testing the possible inflection value $x = 3$, we see that $f''(x) < 0$ and therefore concave down on $(-\infty, 3)$ and for the interval $(3, \infty)$, $f''(x) > 0$ and therefore concave up on $(3, \infty)$.

e. Therefore, 3 is inflection point.

Exercises:

Find absolute extrema for the following functions on closed intervals:

1. $f(x) = x^2 - 2x$ on $[0, 4]$ [minimum $(1, -1)$, maximum $(4, 8)$]

2. $f(x) = 9 - x^2$, on $[-3, 3]$

3. $f(x) = x^2 - 2x + 3$, on $[-4, 1]$

4. $f(x) = 2x/(x^2 + 1)$ on $[-2, 2]$ [minimum $(-1, -1)$, maximum $(1, 1)$]

5. $f(x) = 3x - x^3$, on $[-1, 1]$

6. $f(x) = x^2(x - 1)$, on $[0, 3]$

7. $f(x) = x/(x - 2)$ on $[3, 5]$ [minimum (5, 5/3), maximum (3, 3)]

8. $f(x) = 4x^2 - 2x^4$, on $[-2, 2]$

9. The Profit function for producing x units per week is $P = 160 - 96x + 18x^2 - x^3$. Find maximum profit on the interval $[0, 10]$.

10. The sum of two non–negative numbers is 10. Find the maximum product.

BUSINESS APPLICATIONS:

Three basic functions of importance to a manufacturer are:

1. Cost function
2. Revenue function
3. Profit function

$C(x)$ = cost of producing x units during a time period
$R(x)$ = revenue from selling x units during a time period
$P(x)$ = profit obtained by selling x units during a time period
Profit = Revenue − Cost, *i.e.*, $P(x) = R(x) − C(x)$

The total cost $C(x)$ of producing x units can be expressed as a sum $C(x) = a + M(x)$, where a is a constant called *overhead* and $M(x)$ is function representing *manufacturing cost* of producing x units.

The overhead, which includes such fixed costs as rent and salaries, does not depend on x; it must be paid even if nothing is produced. The manufacturing cost $M(x)$, which includes such items as cost of materials, labor, and energy use, depends on the number x of units produced. If a company can sell all the items it produced for p dollars each, then its total revenue $R(x)$ will be $R(x) = p(x)$ and its profit $P(x)$ will be $P(x) = R(x) − C(x)$ which is $p(x) − C(x)$.

Depending on many factors, such as the number of employees, available manufacturing technology, economic conditions, etc., there will be some upper limit L on the number of units that the manufacturer is capable of producing and selling. By determining the value of x over the interval [0, L] that maximizes profit, the company can determine how many units must be manufactured and sold to produce the greatest profit.

Example:

A medicine manufactured by a pharmaceutical company is sold for $200 per unit. The cost $C(x)$ of producing x units is given by a formula $C(x) = 500000 + 80x + 0.003x^2$. The company can make at most 30000 units in a given time period. How many units of this medicine must be made and sold to maximize profit?

Revenue $R(x) = 200x$
Profit $P(x) = R(x) − C(x) = 200x − (500000 + 80x + 0.003x^2)$

Since the production capacity is at most 30000 units, x is in the interval [0, 30000]. We need to find Absolute Maximum of $P(x) = 200x - (500000 + 80x + 0.003x^2)$.

1. Find critical values of P(x) by setting the derivative of P(x) to zero and solving for x

2. Evaluate the function at the critical points *and* the endpoints of the interval [0, 30000]

The largest value of P(x) is absolute maximum of over the interval [0, 30000], which is the maximum profit.

$dP/dx = 200 - (80 + 0.006x) = 120 - 0.006x$. If $dP/dx = 0$ then $120 - 0.006x = 0$, thus $120 = 0.006x$ and $x = 20000$. The number of units $x = 20000$ is within the interval [0, 30000] so the maximum profit must occur at one of three points:

$x = 0$ (endpoint)
$x = 20000$ (point at which the derivative is equal to zero)
$x = 30000$ (endpoint)

Substituting these values into $P(x) = 200x - (500000 + 80x + 0.003x^2)$ we obtain:

If $x = 0$, $P(x) = -500000$
If $x = 20000$, $P(x) = 700000$
If $x = 30000$, $P(x) = 400000$

Therefore, producing 20000 units of medication will produce the greatest profit of $700,000.

Example:

Find the level of output at which profit is maximized if revenue is $R = 600x - 5x^2$ and the cost is $C = 20x + 320$.

The profit function is $P = R - C = (600x - 5x^2) - (20x + 320)$
$P = -5x^2 + 580x + 320$. The derivative of P(x) is $d/dx [P(x)] = -10x + 580$.

To find critical values we need to set the derivative equal to zero and solve for x:
$-10x + 580 = 0$, then $-10x = -580$, and $x = 58$.

To test for concavity and relative extrema we need the second derivative:
d/dx $[-10x + 580] = -10$. Since the second derivative is negative for all x, the profit function P is concave down at the critical value x = 58 and so the maximum profit occurs at this point: $P(58) = -5(58)^2 + 580(58) + 320 = 16,500$ units of output.

Example:

The price of some product is p, the revenue is $R(x) = -0.05p^2 + 9p + 18$. What unit price maximizes the revenue?

The derivative of R(x) is $2(-0.05)p + 9 = -0.1p + 9$. Setting the derivative equal to zero and solving for p, we obtain: $-0.1p + 9 = 0$, then $-0.1p = -9$, and so $p = -9/(-0.1) = 90$.

The critical value p = 90 may be the maximum of function R(x). Since we do not know the interval for p, we cannot use the procedure from the previous example. Using Second Derivative Test we can verify if 90 is indeed the maximum: Second derivative is $d/dx[-0.1p + 90] = -0.1$. It is negative for all p, in particular for p = 90. Therefore, the revenue function is always concave down. Thus, the revenue is maximized when the price is $90.

Marginals:

The derivatives of Profit, Revenue and Cost functions are called Marginal Profit, Marginal Revenue and Marginal Cost. They are the *additional* Profit, Revenue and Cost that result from producing and selling one additional unit after x units have been sold. For example, after producing and selling 1000 games, what additional revenue, if any, will result from producing and selling 1001st game?

One of the basic principles is that the maximum profit occurs at a point where the cost of producing and selling an additional unit is exactly equal to the revenue generated by the sale of this additional unit. The profit function is maximized when MR is equal to MC.

Marginal Revenue is the derivative of Revenue: $MR = d/dx[R(x)]$
Marginal Cost is the derivative of Cost: $MC = d/dx[C(x)]$

Example:

The revenue of selling x units of product is given by $R(x) = 5x^2 + 10x + 60$, and cost is $C(x) = 2000x + 20$. How many units must be sold to maximize profit?

Profit is maximized when marginal revenue equals marginal cost, that is, MR = MC, where MR is the derivative of revenue and MC is the derivative of cost.

$MR = d/dx[5x^2 + 10x + 60] = 5(2x) + 10 = 10x + 10$
$MC = d/dx[2000x + 20] = 2000$

Therefore, MR = MC means that

$10x + 10 = 2000$
$10x = 2000 - 10$
$10x = 1990$
$x = 1990/10 = 199$

Maximize profit by selling 199 units.

Exercises:

1. The profit P for selling x units of a product is $P(x) = -x^3 + 45x^2 + 1200x + 80$ for maximum of 50 units per month. What amount of sales produces maximum profit per month?

2. The revenue for selling x units is $R(x) = -0.01x2 + 5x$. Find Marginal Revenue for 20 units.

3. The revenue for selling x units of a product is $R(x) = -0.01x^2 + 2x + 2000$, and the cost is $C(x) = 0.08x + 1000$. Find marginal revenue, marginal cost, and marginal profit for a production *after* 55 units.

4. The cost of producing x units of a product is $C(x) = 0.004x^2 - 9.6x + 7840$. How many units should be produced to minimize the cost?

CHAPTER 4 – INTEGRALS

INDEFINITE INTEGRAL – INTEGRATION AS ANTI–DIFFERENTIATION:

We can think of Integration as the opposite of Differentiation. When differentiating, we were given a function computed its derivative. Integration allows us to construct the original function *from* its derivative.

For example, suppose that the derivative of f(x) is 2x. What function has a derivative 2x? One such function is x^2, another is $x^2 + 3$ or $x^2 - 7$ or $x^2 + 345$. Actually, it is $x^2 + C$, where C is any real number. Because the derivative of any number is zero, it does not matter what number we add to the variable x^2 because derivative of any real number zero.

The Anti–derivative of A is B if and only if the derivative of B is A.

For example, the integral $\int 3x^2 \, dx = x^3 + C$ because the derivative of $x^3 = 3x^2$. The Integral of function f(x) is written $\int f(x) \, dx$.

Integration Rules:

1. $\int c \cdot f(x) \, dx = c \cdot \int f(x) \, dx$, c is a real number
2. $\int [f(x) + g(x)] \, dx = \int f(x) \, dx + \int g(x) \, dx$
3. $\int [f(x) - g(x)] \, dx = \int f(x) \, dx - \int g(x) \, dx$
4. $\int x^n \, dx = (x^{n+1})/(n + 1) + C$, except for $n = -1$.

Note: If $n = -1$ then $\int x^{-1} \, dx = \ln |x| + C$, which is a natural logarithm of the absolute value of x, plus a constant.

Examples:

$\int x^8 \, dx = (x^9)/9 + C$

$\int x^{20} \, dx = (x^{21})/21 + C$

$\int x^{-6} \, dx = (x^{-6+1})/(-6 + 1) + C = (x^{-5})/(-5) + C$

$\int x^{1/2} \, dx = (x^{1/2+1})/(1/2 + 1) + C = (x^{3/2})/(3/2) + C = (2/3)(x^{3/2}) + C$

Exercises:

Find the following Integrals using the rules of Integration:

1. $\int 4x^4 \, dx =$

2. $\int (3x + 4x^3) \, dx =$

3. $\int (x^5 - x^2) \, dx = x^6/6 - x^3/3 + C$

4. $\int 3x^2(x + 1) \, dx =$

5. $\int 5(x + 3)^2 \, dx =$

6. $\int 4x(2x^2 - 1)^5 \, dx = (1/6)((2x^2 - 1)^6 + C$

7. $\int (x^3 - 6x + 2) \, dx =$

8. $\int (x^{2/3} - x^{-1/3}) \, dx =$

9. $\int (x - 5)^{1/3} \, dx = (3/4)(x - 5)^{4/3} + C$

10. $\int (x^2 + 1)^2 \, dx =$

INTEGRATION BY SUBSTITUTION:

If an integral is expressed in the form of function with exponent *times* the derivative of this function, *i.e.*, $\int f(x)^n [f(x)]'\, dx$, then this integral may be solved by substitution where $f(x)$ may be called u and then $[f(x)]'$ is called du:
$\int f(x)^n [f(x)]'\, dx = \int u^n\, du$.

For example, for the integral $\int (x^2)(2x)\, dx$, x^2 is u and $2x$ is du because the derivative of x^2 is $2x$.

If $f(x) = u$, then $\int f(x)^n [f(x)]' = \int u^n\, du = (u^{n+1})/(n+1) + C$.

Example:

Integrate $\int (x^2 + 5)^2 (2x)\, dx$.

Let $u = x^2 + 5$ then $du = 2x$. Therefore, $\int (x^2 + 5)^2 (2x)\, dx = \int u^2\, du = u^3/3 + C$.
Substituting $x^2 + 5$ instead of u we obtain $(x^2 + 5)^3/3 = (x^2 + 5)^3/3 + C$.
Thus, $\int (x^2 + 5)^2 (2x) = (x^2 + 5)^3/3 + C$.

Example:

Integrate $\int (2x)(x^2 + 1)^{10}\, dx$.

Let $u = x^2 + 1$ then $du = 2x$. Therefore, $\int (2x)(x^2 + 1)^{10} = \int u^{10}\, du = u^{11}/11 + C$.
Substituting $x^2 + 1$ instead of u we obtain $(x^2 + 1)^{11}/11 + C$.
Thus, $\int (2x)(x^2 + 1)^{10}\, dx = (x^2 + 1)^{11}/11 + C$.

Note: the integral is in the form $(2x)(x^2 + 1)^n$ where $2x$ is the derivative of $(x^2 + 1)$.

Example:

Integrate $\int (x^4 - 1)^2 (4x^3)\, dx$.

Let $u = x^4 - 1$ then $du = 4x^3$. Therefore, $\int (x^4 - 1)^2 (4x^3)\, dx = \int u^2\, du + C$, which is $u^3/3 + C$. Substituting $(x^4 - 1)$ instead of u we obtain $(x^4 - 1)^3/3 + C$.
Thus, $\int (x^4 - 1)^2 (4x^3)\, dx = (x^4 - 1)^3/3 + C$.

Example:

Integrate $\int (x^2 + 3x + 5)^8 (2x + 3) \, dx$.

Let $u = x^2 + 3x + 5$, then $du = 2x + 3$.
Then $\int (x^2 + 3x + 5)^8 (2x + 3) \, dx = \int u^8 \, du = (u^{8+1})/(8+1) + C = (u^9)/9 + C$.
Since $u = x^2 + 3x + 5$, $(u^9)/9 + C = (x^2 + 3x + 5)^9/9 + C$.

Example:

Integrate $\int (x^3 + 7)^5 (3x^2) \, dx$.

Let $u = x^3 + 7$, then $du = 3x^2$.
Then $\int (x^3 + 7)^5 (3x^2) \, dx = \int u^5 \, du = (u^{5+1})/(5+1) + C = (u^6)/6 + C$.
Thus, $\int (x^3 + 7)^5 (3x^2) \, dx = (u^6)/6 + C = (1/6)u^6 = (1/6)(x^3 + 7)^6 + C$.

Example:

Integrate $\int (4x^3 - 6x)(x^4 - 3x^2 + 4)^3 \, dx$.

Let $u = x^4 - 3x^2 + 4$ then $du = 4x^3 - 6x$.
Then $\int (4x^3 - 6x)(x^4 - 3x^2 + 4)^3 \, dx = \int u^3 \, du = (u^{3+1})/(3+1) + C = (u^4)/4 + C$.
Thus, $\int (4x^3 - 6x)(x^4 - 3x^2 + 4)^3 \, dx = (1/4)u^4 = (1/4)(x^4 - 3x^2 + 4)^4 + C$.

Example:

Integrate $\int 5(5x - 1)^{-3} \, dx$.

Let $u = (5x - 1)$ then $du = 5$.
Then $\int 5(5x - 1)^{-3} \, dx = \int u^{-3} \, du = (u^{-3+1})/(-3+1) + C = (u^{-2})/(-2) + C =$
$= (-1/2)(5x - 1)^{-2} + C$.

Examples:

$\int 6(6x + 1)^4 \, dx = (1/5)(6x + 1)^5 + C$, because the derivative of $(6x + 1)$ is 6.

$\int (3 - 4x^2)^{1/3} (-8x) \, dx = (3/4)(3 - 4x^2)^{4/3} + C$, because the derivative of $(3 - 4x^2)$ is $(-8x)$.

Example:

Find y if $dy/dx = 10x^2/(1 + x^3)^{1/2}$. The value of y is the integral of dy/dx, i.e., y is antiderivative of $10x^2/(1 + x^3)^{1/2}$. Let's re-write the Integral without division:
$\int 10x^2(1 + x^3)^{-1/2}$

We would like to use substitution to solve $\int 10x^2(1 + x^3)^{-1/2}$, but derivative of x^3 is $3x^2$; we do not have $3x^2$, we only have x^2; thus, we need to multiply by 3 to "create" $3x^2$. At the same time, we must divide by 3 to preserve the original equation. We get: $(10/3)\int 3x^2(1 + x^3)^{-1/2}$. Now we can use integration by substitution:

$y = (10/3)\int(1 + x^3)^{-1/2}(3x^2)\,dx = (10/3)(2)[(1 + x^3)^{1/2}] = (20/3)(1 + x^3)^{1/2} + C.$

Exercises:

1. $\int 6(6x + 1)^4\,dx = (6x + 1)^5/5 + C$

2. $\int 2x(x^2 + 8)\,dx =$

3. $\int 16x(8x^2 - 9)^3\,dx =$

4. $\int x^3(x^4 + 3)^2\,dx = (x^4 + 3)^3/12 + C$

5. $\int 3x^2(x^3 - 1)\,dx =$

6. $\int 5x^4(x^5 + 6)^4\,dx =$

7. $\int 12x(6x^2 + 8)^{1/2}\,dx = (2/3)(6x^2 + 8)^{3/2} + C$

8. $\int 2x(x^2 + 1)^{1/2}\,dx =$

9. $\int 5x^4(x^5 + 9)^{-1/2}\,dx =$

10. $\int(-3/x^4)(x^{-3} - 3)^3\,dx = (1/4)(x^{-3} - 3)^4 + C$

INITIAL CONDITIONS AND PARTICULAR SOLUTIONS TO DIFFERENTIAL EQUATIONS

Generally, we can find almost any function given its derivative. Given the derivative of function, we simply integrate this derivative to obtain the original function.

Antiderivative of A is B if and only if the derivative of B is A. For example, the integral $\int 3x^2 \, dx$ is equal to $x^3 + C$ because $d/dx[x^3] = 3x^2$. Therefore, we can use Integration to construct the original function *from* its derivative, provided we can find the value of a constant C.

An equation of the type $dy/dx = f(x)$ is called *differential equation* because it involves the derivative of function $f(x)$. For example, if $dy/dx = x^2$ then the solution to this differential equation is $y = \int x^2 \, dx = x^3/3 + C$, where C is a constant. Graphically, it means that the solution set consists of the vertical translations of the curve of $x^3/3$, each translation being a particular solution corresponding to a given value of C. To find a *particular* solution, we need to know the value of $y = f(x)$ for some x. The value of y for a given x is called *initial condition*.

Example:

Find a particular solution of the differential equation $dy/dx = (x)^{1/3}$ subject to initial condition $y(1) = 2$, that is when $x = 1$, $y = 2$.

$y(x) = \int x^{1/3} \, dx = (3/4)(x)^{4/3} + C$. Since $y(1) = 2$, $(3/4)(1)^{4/3} + C = 2$, and so $C = 5/4$. Therefore, the solution is $y(x) = (3/4)(x)^{4/3} + 5/4$.

Example:

Find the equation of the curve with slope $m = 2x + 1$ of the tangent line that passes through point $(-3, 0)$.

The differential equation is $dy/dx = 2x + 1$. Thus, $y(x) = \int (2x + 1) \, dx = x^2 + x + C$. Because $y = 0$ when $x = -3$, we have $x^2 + x + C = (-3)^2 + (-3) + C = 0$. Therefore, $C = -6$ and the solution is $y(x) = x^2 + x - 6$.

It means that $y(x) = x^2 + x - 6$ is the equation of the curve with slope $m = 2x + 1$ of the tangent line that passes through $(-3, 0)$.

Example:

Find the Cost function if the Marginal Cost function is MC = 0.08x + 1, where x is the number of units produced.

We can find Cost function given Marginal Cost function because Marginal Cost is the derivative of Cost, therefore Cost function is the integral of the Marginal Cost function: ∫MC dx = Cost + C. Cost = ∫MC dx = ∫(0.08x + 1) dx = 0.04x² + x + C.

If we know the cost of producing a given number of units, we can compute the constant of integration. Let's say that it costs $6800 to produce 200 units. Substituting these numbers into the Cost function, we obtain

6800 = 0.04 (200²) + 200 + constant
6800 = 1800 + constant
5000 = constant
Therefore, the Cost function is 0.04x² + x + 5000.

Example:

Find Distance function if Instantaneous Velocity of an object is V(t) = 14t + 3 feet per second at some instant of time t.

We can find distance function given Velocity function because Velocity is the derivative of distance function: V = d/dx[D]. Therefore, integral of Velocity is distance function plus a constant: ∫V dx = D + constant.
Therefore, ∫V dt = D + C. Thus, distance function is D = ∫14t + 3 dx = 7t² + 3t + C.

We could find a particular solution if we were given value of D for a particular time t.

Example:

What is the function f(x) if the slope of a line tangent to f(x) at the point (1, − 3) is given by the formula 12x³ − 6x + 5?

The formula 12x³ − 6x + 5 is the derivative of the original function f(x). Therefore, the original function f(x) is the integral of the formula:

f(x) = ∫12x³ − 6x + 5 dx = 12(x⁴)/(4) − 6(x²)/(2) + 5x + C = 3x⁴ − 3x² + 5x + C

We can find the constant C by evaluating this expression at the given point $(1, -3)$, therefore, $f(1) = 3(1)^4 - 3(1)^2 + 5(1) + C = -3$ and therefore, $3 - 3 + 5 + C = -3$.

Therefore, $C = -8$ and the original function is $f(x) = 3x^4 - 3x^2 + 5x - 8$.

Exercises:

1. Find particular solution of $f'(x) = 8x^3 + 5$, where $f(1) = -4$. $[C = -11]$

2. Find the general form of the cost function if the marginal cost function is $m(x) = 0.04x^2 + 20x - 120$, where x is the number of units produced.

3. Find position (distance) function if the velocity is $V = 14t + 3$. $[7t^2 + 3t]$

4. Find function $f(x)$ if the slope of the line tangent to this function at the point $(1, 3)$ is given by the formula slope $= 4x^2 - x$.

5. The derivative of function is $f'(x) = 1 + (\sec x)(\tan x)$.
Find the original function $f(x)$ if $f(\pi) = 2$. $[f(x) = x + \sec x + 3 - \pi]$.

DEFINITE INTEGRALS – AREA UNDER THE GRAPH:

The integrals that we dealt with so far are called *indefinite* integrals. Indefinite integral is an algebraic expression. The *definite* integral is a number.

This number can be thought of as the area under the graph of function f(x) and above x- axis, between two values of x: $x = a$ and $x = b$.

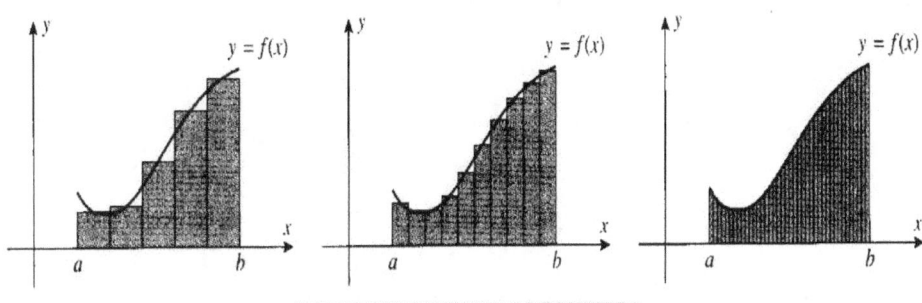

Fundamental Theorem of Calculus:

Let f(x) denote the derivative of F(x): $d/dx[F(x)] = f(x)$.
If f(x) is continuous over the closed interval from x_1 to x_2 then the integral of f(x) from x_1 to x_2 is equal to $F(x_2) - F(x_1)$:

$$\int f(x)\, dx = F(x_2) - F(x_1)$$

It simply means that to evaluate integral $\int f(x)\, dx$, we need to do three things:

1. Find the antiderivative F(x)
2. Evaluate the antiderivative at x_1 and x_2, obtaining two numbers: $F(x_1)$ and $F(x_2)$
3. Subtract the first number from the second: $F(x_2) - F(x_1)$

Example:

Evaluate $\int 3x^2\, dx$ from $x = 2$ to $x = 5$.

First of all, $\int 3x^2\, dx = 3\int x^2\, dx$ (rule #1)
$\int x^2\, dx = (x^3)/3\, dx + C = (1/3)x^3$ (rule #4)

Therefore, $\int 3x^2\, dx = 3(1/3)x^3 + C = x^3 + C$.

Thinking in terms of anti-differentiation we can see that the integral of $3x^2$ is x^3 simply because the derivative of x^3 is $3x^2$. Now we evaluate this antiderivative at the limits of integration:

Evaluating x^3 at $x = 5$, we obtain $5^3 = 125$
Evaluating x^3 at $x = 2$, we obtain $2^3 = 8$
Subtracting: $125 - 8 = 117$.

Think of this number 117 as the area under the graph of $3x^2$ from $x = 2$ to $x = 5$:

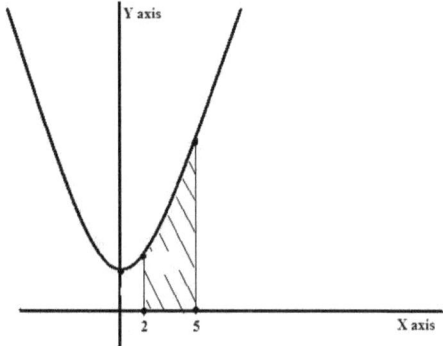

Example:

Find the area under the graph of $y = 2x - 1$ from $x = 1$ to $x = 4$.

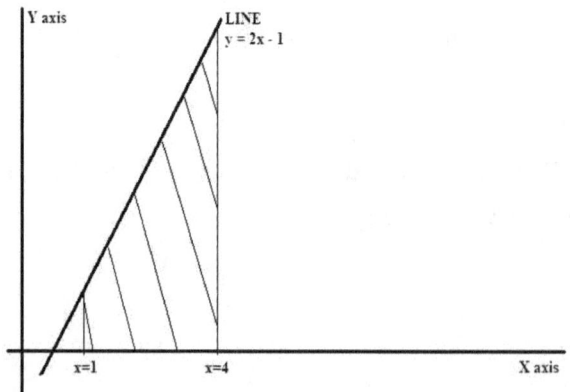

The integral of $f(x) = 2x - 1$ is $\int(2x - 1)\,dx = \int 2x\,dx - \int 1\,dx = x^2 - x$.

Evaluating $(x^2 - x)$ at $x = 4$ and $x = 1$ and subtracting we obtain
$(4^2 - 4) - (1^2 - 1) = (16 - 4) - (0) = 12$.

Note: we can also evaluate the area in this particular example by purely geometric means because the function f(x) = 2x − 1 is a straight line so that the area under the graph is a trapezoid whose area is one half of the sum of parallel sides times the height. The height of this trapezoid is the distance between x = 4 and x = 1. The lengths of parallel sides are obtained by substituting x = 1 and x = 4 into the function y = 2x − 1, obtaining 1 and 7. Thus, the area is (1/2)(1 + 7)(3) = 12.

Example:

Find the area under the graph of $f(x) = x^2 + 2$ from x = 1 to x = 3.

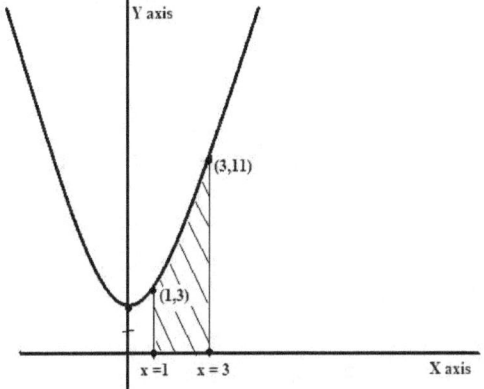

We need to integrate the function $f(x) = x^2 + 2$ using the rules of integration, which means integrating x^2 and 2 separately (rule #2).

∫(x² + 2) dx = ∫x² dx + ∫2 dx,

∫x² dx = (1/3)x³ (rule #4)

∫2 dx = 2x (rule #1 and #4)

Therefore ∫(x² + 2) dx = (1/3)x³ + 2x

Evaluating (1/3)x³ + 2x at x = 3 we obtain (1/3)(3)³ + 2(3) = 15
Evaluating (1/3)x³ + 2x at x = 1 we obtain (1/3)(1)³ + 2(1) = 7/3
Subtracting, we obtain 15 − 7/3 = 38/3.

The area under the graph of $f(x) = x^2 + 2$ from x = 1 to x = 3 is 38/3 square units.

Example:

Find the area above x–axis and under the graph of $f(x) = 2 + x - x^2$.

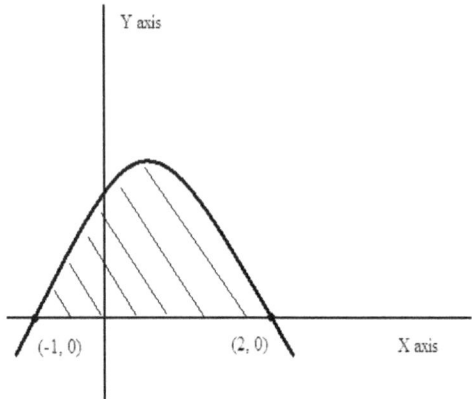

First, we need to find the points of intersection of the graph of f(x) and the x-axis. These are the points where $f(x) = 2 + x - x^2$ is equal to zero.

$f(x) = 2 + x - x^2 = (2 - x)(1 + x) = 0$, thus x = 2 or x = − 1

Therefore, we need to find the area under $f(x) = 2 + x - x^2$ from x = − 1 to x = 2

$\int (2 + x - x^2)\, dx = 2x + (x^2)/2 - (x^3)/3$

Evaluating $2x + (x^2)/2 - (x^3)/3$ at x = 2 we obtain
$2(2) + (2^2)/2 - (2^3)/3 = 4 + 2 - 8/3 = 6 - 8/3 = 10/3$

Evaluating $2x + (x^2)/2 - (x^3)/3$ at x = − 1 we obtain
$2(-1) + (-1^2)/2 - (-1^3)/3 = -2 + 1/2 + 1/3 = -7/6$

Subtracting, we obtain 10/3 − 7/6 = 9/2. The area above x–axis and under the graph of $f(x) = 2 + x - x^2$ is 9/2.

Exercises:

Find the area under the graph of the following functions:

1. $f(x) = 2x + 3$ from 0 to 4 [28]

2. $f(x) = 3x - 6$, from 2 to 4.

3. $f(x) = x^2 - 6$, from 1 to 5.

4. $f(x) = 4x^3 - 6$ from 1 to 2 [9]

5. $f(x) = x^3 - 4x + 1$, from 0 to 3.

6. $f(x) = -x^2 + x$, from 0 to 5.

7. $f(x) = 4x/(2x^2 + 1)^2$ from 2 to 5 [14/153]

8. $f(x) = 3x^2 - 2x + 1$, from 1 to 2.

9. $f(x) = 4x^3 - 6$, from -1 to 3.

10. $f(x) = x^2 + 8x - 1$, from 2 to 3.

AREA BETWEEN TWO GRAPHS:

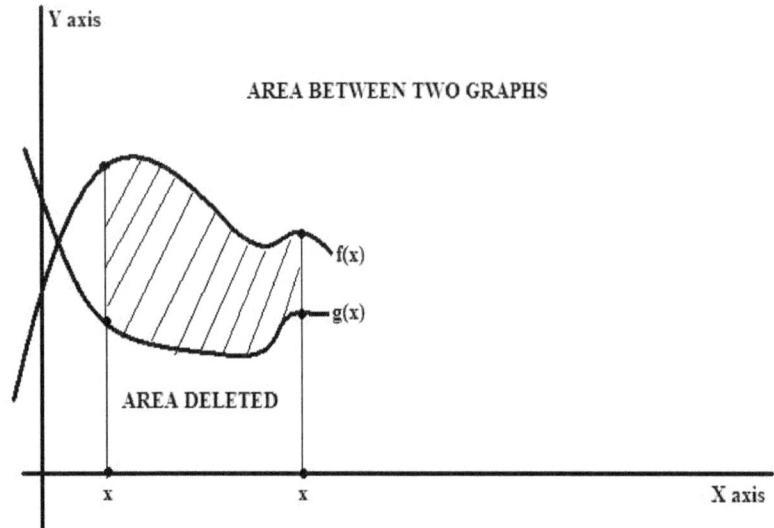

The area between two graphs is simply the area under the top graph with the area under the bottom graph deleted: Area = ∫(top graph − bottom graph) dx, evaluated between x_1 and x_2.

Example:

What is the area between the line $f(x) = -x + 7$ and parabola $f(x) = x^2 - 6x + 11$, from $x_1 = 1$ to $x_2 = 4$?

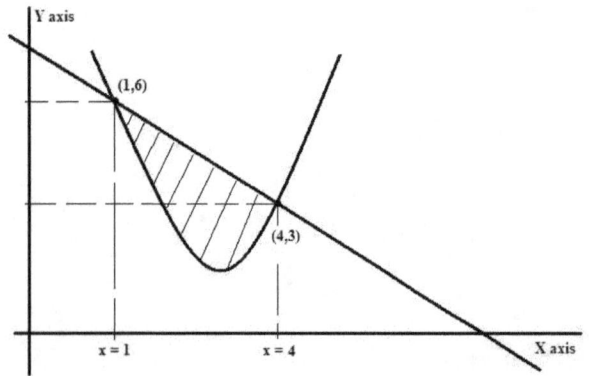

Area = $\int[(-x + 7) - (x^2 - 6x + 11)]\,dx = \int -x^2 + 5x - 4\,dx$.

By integrating (that is, finding anti-derivatives of $(-x^2)$, $(5x)$, and (-4) we obtain:

$\int (-x)^2\,dx = (-1/3)(x^3)$, $\int 5x\,dx = 5/2(x^2)$, and $\int(-4)\,dx = -4x$

Therefore, $\int -x^2 + 5x - 4 \, dx = (-1/3)(x^3) + (5/2)(x^2) - 4x$

Evaluating this expression at x = 4: $-1/3(4^3) + 5/2(4^2) - 4(4) = 8/3$
Evaluating this expression at x = 1: $-1/3(1^3) + 5/2(1^2) - 4(1) = -11/6$

Subtracting: $[(-1/3)(4^3) + (5/2)(4^2) - 4(4)] - [(-1/3)(1^3) + (5/2)(1^2) - 4(1)] =$
$= 8/3 - (-11/6) = 9/2$ square units.

This is the area between the line $f(x) = -x + 7$ and the parabola $f(x) = x^2 - 6x + 11$, from $x_1 = 1$ to $x_2 = 4$.

Example:

Find the area between the graphs of $f(x) = -x^2 + 4x - 3$ and $g(x) = x^2 - 4x + 3$ from x = 1 to x = 3.

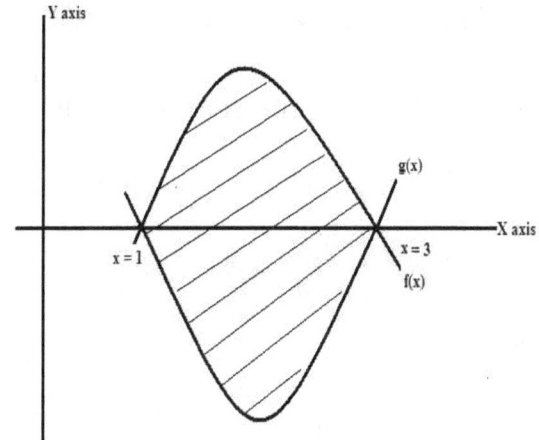

Area = $\int$[top curve − bottom curve] dx, evaluated from 1 to 3:

$\int [-x^2 + 4x - 3] - [x^2 - 4x + 3] \, dx = \int [-x^2 + 4x - 3 - x^2 + 4x - 3] \, dx =$

$= \int -2x^2 + 8x - 6 \, dx = (-2/3)(x^3) + 4x^2 - 6x$

Evaluating this expression at x = 3: $(-2/3)(3^3) + 4(3^2) - 6(3) = 0$
Evaluating this expression at x = 1: $(-2/3)(1^3) + 4(1^2) - 6(1) = -8/3$

Subtracting we obtain the area: $0 - (-8/3) = 8/3$ square units.
This is the area between the graphs of $f(x) = -x^2 + 4x - 3$ and $g(x) = x^2 - 4x + 3$.

Example:

Find the area between the graphs of $f(x) = x^4 + 1$ and $g(x) = 2x^2$.

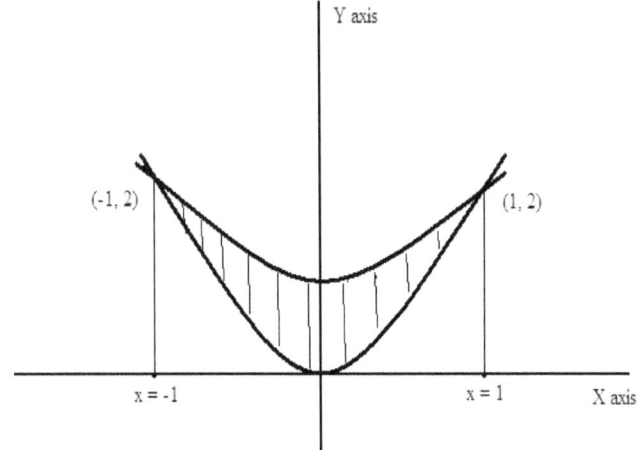

Find the points of intersection of $f(x) = x^4 + 1$ and $g(x) = 2x^2$ where $f(x) = g(x)$.

If $f(x) = g(x)$ then $x^4 + 1 = 2x^2$ or $2x^2 - x^4 + 1 = (x^2 - 1)^2 = 0$, thus $x^2 - 1 = 0$ and $x = 1$ or $x = -1$.

Find $f(1)$ and $g(1)$:

evaluating at $x = 1$, $f(1) = 1^4 + 1 = 2 = 2(1)^2 = g(1)$
evaluating at $x = -1$, $f(-1) = (-1)^4 + 1 = 2 = 2(-1)^2 = g(-1)$

Therefore, the points of intersection are $(-1, 2)$ and $(1, 2)$.

Area = $\int$[top curve − bottom curve] dx, evaluated from -1 to 1:
$\int[(x^4 + 1) - (2x^2)]\,dx = \int x^4 - 2x^2 + 1\,dx = (x^5)/5 - 2(x^3)/3 + x$

Evaluating this expression at $x = 1$, $x = -1$, and then subtracting, we obtain the area: $[(1^5)/5 - 2(1^3)/3 + (1)] - [(-1^5)/5 - 2(-1^3)/3 + (-1)] = 16/15$ square units.

Example:

Find the area between the graphs of $f(x) = x^3 - 4x$ and $g(x) = 5x$.

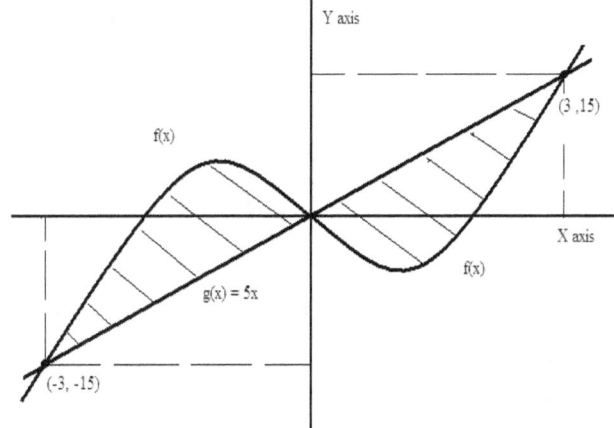

We must find the points of intersection of $f(x) = x^3 - 4x$ and $g(x) = 5x$.
If $f(x) = g(x)$ then $x^3 - 4x - 5x = x^3 - 9x = x(x^2 - 9) = 0$, thus $x = -3, x = 0, x = 3$.

evaluating at $x = -3$, $f(-3) = (-3^3) - 4(-3) = -15$
evaluating at $x = 0$, $f(0) = (0^3) - 4(0) = 0$
evaluating at $x = 3$, $f(3) = (3^3) - 4(3) = 15$

Therefore, the points of intersection are $(-3, -15)$, $(0, 0)$, and $(3, 15)$.
The area consists of two regions: from $x = -3$ to $x = 0$ and from $x = 0$ to $x = 3$.

Test arbitrary points x in each interval.

On the interval $[-3, 0]$ function $f(x) = x^3 - 4x$ is higher than the function $g(x) = 5x$, therefore the area of this region is:
$\int [(x^3 - 4x) - 5x] \, dx = \int x^3 - 9x \, dx = (1/4)(x^4) - (9/2)(x^2)$

Evaluating this expression at $x = 0$, $x = -3$, and then subtracting, we obtain the area of the region between $x = -3$ and $x = 0$:
$[(1/4)(0^4) - (9/2)(0^2)] - [(1/4)(-3^4) - (9/2)(-3^2)] = 81/4$

On the interval $[0, 3]$ the function $f(x) = x^3 - 4x$ is lower than the function $g(x) = 5x$, therefore the area of this region is
$\int [5x - (x^3 - 4x)] \, dx = \int -x^3 + 9x \, dx = (9/2)(x^2) - (1/4)(x^4)$

Evaluating this expression at $x = 3$, $x = 0$, and then subtracting, we obtain the area of the region between $x = 0$ and $x = 3$: $[(9/2)(3^2) - (1/4)(3^4)] - 0 = 81/4$.

The area between the graphs of $x^3 - 4x$ and $5x$ is the area of the region between $x = -3$ and $x = 0$ plus the area of the region between $x = 0$ and $x = 3$, which is $81/4 + 81/4 = 81/2$.

Exercises:

Find the areas between the graphs of

1. $f(x) = x/2 + 2$ and $g(x) = x^2$ between $x = 0$ and $x = 1$ [23/12]

2. $f(x) = x^3 + 2x^2$ and $g(x) = 3x$

3. $f(x) = x^2$ and $y = x + 2$ [9/2]

4. $f(x) = x^2$ and $g(x) = x + 6$

5. $f(x) = x^3$ and $g(x) = 2x$ between $x = 0$ and $x = 1$ [3/4]

VOLUME OF A SOLID OF REVOLUTION:

If the function f(x) is continuous on the interval (x_1, x_2) and its graph from x_1 to x_2 is rotated around the x-axis, then the volume of the resulting solid is given by the integral $\int \pi [f(x)]^2 \, dx$ evaluated from x_1 to x_2.

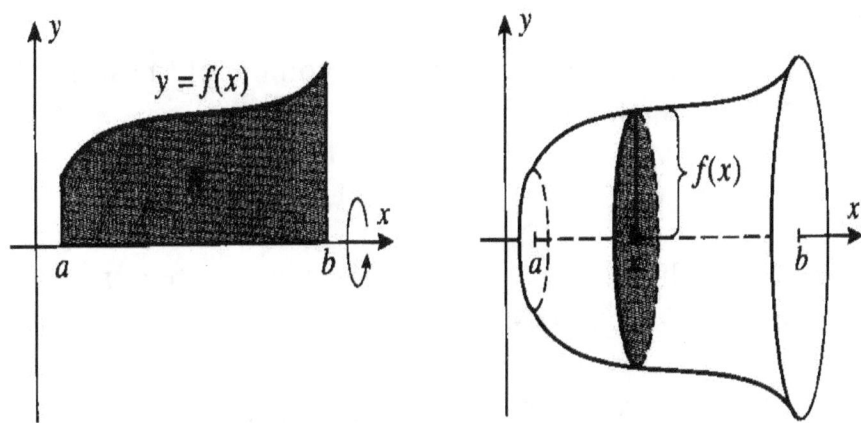

Example:

The function $f(x) = x^2$ from $x = 0$ to $x = 1$ is rotated around the x-axis. What is the volume of the resulting solid?

Volume = $\int \pi [x^2]^2 \, dx = \int \pi (x^4) \, dx = \pi \int x^4 \, dx = \pi (x^5)/5 = \pi (x^5)/5$

Evaluating at $x = 1$, $x = 0$ and subtracting we obtain the volume of the solid of revolution: $\pi (1^5)/5 - \pi (0^5)5 = \pi/5$ cubic units.

Example:

The function $f(x) = (2/3)(x)$ from $x = 0$ to $x = 9$ is rotated around the x-axis. What is the volume of the resulting solid?

Volume = $\int \pi [(2/3)(x)]^2 \, dx = \pi \int [(4/9)x^2] \, dx = \pi (4/9) \int x^2 \, dx = \pi (4/9)(x^3/3) =$
$= 4\pi (x^3)/27$

Evaluating at $x = 0$, $x = 9$ and subtracting we obtain the volume of the solid of revolution: $4\pi (9^3)/27 - 4\pi (0^3)/27 = 108\pi$ cubic units.

Example:

Find the volume of the solid of revolution generated by revolving around x-axis the region under the graph of $f(x) = 5x^2$ from $x = 1$ to $x = 3$.

$\int \pi [5x^2]^2 \, dx = \int \pi [25x^4] \, dx = 25\pi \int x^4 \, dx = 25\pi (x^5/5) = 5\pi x^5$

Evaluating at $x = 3$ and $x = 1$ and subtracting we obtain the volume of the solid of revolution: $5\pi(243 - 1) = 1210\pi$ cubic units.

Example:

Find the volume of the solid of revolution generated by revolving the region enclosed by $y = (25 - x^2)^{1/2}$ and $y = 3$ around x-axis.

Points of intersection are: $(25 - x^2)^{1/2} = 3$, so $25 - x^2 - 9 = 0$, and so $x = -4$ or 4.

Volume $= \pi \int [(25 - x^2) - 9] \, dx$. This integral, evaluated from $x = -4$ to $x = 4$, is equal to $256\pi/3$ cubic units.

Example:

Find the volume of the solid of revolution generated by revolving around x-axis the region enclosed by $f(x) = x^2 + \frac{1}{2}$ and $g(x) = x$ over the interval $[0, 2]$.

Volume $= \pi \int ((x^2 + \frac{1}{2})^2 - x^2)^2 \, dx = \pi \int x^4 + 1/4 \, dx = \pi (x/4 + x^5/5)$.
Evaluating this integral from 0 to 2, we obtain volume of $69\pi/10$ cubic units.

Example:

Find the volume of the solid of revolution generated by revolving around x-axis the region enclosed by $x = y^{1/2}$ and $x = y/4$.

If $x = y^{1/2}$ then $y = x^2$, if $x = y/4$ then $y = 4x$.

These graphs intersect when $x^2 = 4x$, $x^2 - 4x = 0$, and so $x = 0$ or $x = 4$.

Volume $= \pi \int (4x)^2 - (x^2)^2 \, dx = \pi \int 16x^2 - x^4 \, dx$. Evaluating this integral from 0 to 4, we obtain volume of $2048\pi/15$ cubic units.

Revolving around y-axis:

When the graph is rotated around x-axis, the function is in terms of x, $y = f(x)$.
If the graph is rotated around y-axis, the function should be in terms of y, $x = g(y)$.

When the graph is rotated around y-axis, the volume of the resulting solid is given by the integral $\int \pi [g(y)]^2 \, dy$, evaluated from y_1 to y_2.

Example:

Find the volume of the solid of revolution generated by revolving around y-axis the region enclosed by $y = x^3$, $x = 0$, and $y = 1$.

We are given the function in terms of x, but we need to represent it in terms of y. Therefore, if $y = x^3$ then $x = y^{1/3}$. This function is enclosed by $y = x^3$, $x = 0$, which is simply y-axis, and the line $y = 1$. Therefore, we integrate from $y = 0$ to $y = 1$.

Volume = $\int \pi [g(y)]^2 \, dy = \int \pi [y^{1/3}]^2 \, dy = \pi \int y^{2/3} \, dy$.
Evaluating the Integral from 0 to 1, we obtain $3\pi/5$ cubic units.

Example:

Find the volume of the solid of revolution generated by revolving around y-axis the region enclosed by $x = (1 + y)^{1/2}$, $x = 0$, and $y = 3$.

This function is bounded by y-axis, the line $y = 3$, and the graph of $x = (1 + y)^{1/2}$. When $x = 0$ then $(1 + y)^{1/2}$ is also equal to zero and so $y = -1$.
Therefore, we integrate from $y = -1$ to $y = 3$.

Volume = $\int \pi [g(y)]^2 \, dy = \int \pi [(1 + y)^{1/2}]^2 \, dy = \pi \int (1 + y) \, dy$.
Evaluating this integral from -1 to 3, we obtain 8π cubic units.

Exercises:

1. Find the volume of the solid of revolution generated by revolving around x-axis the region under the graph of $f(x) = 2x + 1$, from $x = 1$ to $x = 4$. $[117\pi]$

2. Find the volume of the solid of revolution generated by revolving around x-axis the region under the graph of $f(x) = x^2$ from $x = 0$ to $x = 2$.

3. Find the volume of the solid of revolution generated by revolving around x-axis the region bounded by the graphs of $f(x) = x^{1/2}$ and $g(x) = x$. $[\pi/6]$

4. Find the volume of the solid of revolution when the region bounded by the graphs of $f(x) = x^2$ and $g(x) = 4x - x^2$ revolved around x-axis.

5. Find the volume of the solid of revolution generated by revolving around x-axis the region bounded by the graphs of $f(x) = x^2 - 4x$ and $y = 0$. $[512\pi/12]$

6. Find the volume of the solid of revolution generated by revolving around x-axis the region bounded by the graphs $x = y^2 - 4y$ and $x = -y$.

7. Find the volume of the solid of revolution generated by revolving around y-axis the region bounded by the graph of $y = x^2$, between the lines $x = 0$ and $y = 4$. $[8\pi]$

THE END

www.ingramcontent.com/pod-product-compliance
Lightning Source LLC
Chambersburg PA
CBHW080300180526
45167CB00006B/2607